Periplus

Periplus

Poetry in Translation

Edited by

DANIEL WEISSBORT
and
ARVIND KRISHNA MEHROTRA

Contributing Editors

A.K. Ramanujan
Jason Wilson
Grzegorz Musial
Tomislav Longinović

DELHI
OXFORD UNIVERSITY PRESS
OXFORD NEW YORK
1993

Oxford University Press, Walton Street, Oxford OX2 6DP

New York Toronto
Delhi Bombay Calcutta Madras Karachi
Kuala Lumpur Singapore Hong Kong Tokyo
Nairobi Dar es Salaam
Melbourne Auckland
and associates in
Berlin Ibadan

ISBN *0 19 563234 6*

Typeset by Rastrixi, New Delhi 110030
Printed in India at Rekha Printers Pvt. Ltd., New Delhi 110020
and published by Neil O'Brien, Oxford University Press
YMCA Library Building, Jai Singh Road, New Delhi 110001

Contents

Introduction

Periplus, its prefix, *peri-*, cognate with the Sanskrit *pari* (round, around, about), is 'The action of sailing round, a circumnavigation; a voyage (or journey) round a coastline, etc.'; in the transferred sense, as in *The Periplus of the Erythraean Sea*, a first-century account that describes the coastal routes from Egypt to India, and along East Africa, it is 'A narrative of such a voyage' (OED).

Periplus: Poetry in Translation attempts to do both. It sails round or along certain coastlines, some of which might be unfamiliar, and whenever possible it will narrate what occurred or did not occur on the journey. A degree of randomness is inevitable in a volume not restricted to any one period or country, and the translation voyages undertaken even here range from A.K. Ramanujan reworking from existing materials the poems of the fourteenth-century Kashmiri saint-poet Lalla, to Weissbort and Musial collaborating to bring over into English the experimental later work of Miron Bialoszewski; and from a Korean shaman song to Arun Kolatkar's Marathi poems, translated by himself in response to various reading invitations, here gathered for the first time. The volume also contains translations of Attila József (Hungary), Grzegorz Musial and Katarzyna Borun-Jagodzinska (Poland), Toshika Fujioka (Japan), and of two Hindi poets, Kunwar Narain and Mangalesh Dabral; essays by Marvin Bell and Tomislav Longinović; an interview with Joseph Brodsky; a review by Jason Wilson of recent Spanish American and Spanish poetry in English translation; and a section on James S. Holmes, the Amsterdam-based American poet, translator, and theorist who died in 1986.

In the Introduction to *Modern Poetry in Translation: 1983*, Ted Hughes, looking back on 'the unique tidal wave of poetry translation that swept through English (and through the other chief Western languages) in the sixties and the early seventies', said that many of the translations done at the time now form 'a great series of naturalised poetic monuments'. That tidal wave, at whose crest stood poets like Miroslav Holub, Zbigniew Herbert, Vasko Popa, Janos Pilinszky, and Yehuda Amichai, touched cities on the Erythreaean Sea also, and if you were a young poet in Bombay in the 1960s, the chances are that you'd be reading 'Zito the Magician' and 'Elegy of Fortinbras' rather than 'Church Going'. The wave, however, was a wholly European affair.

MPT was founded in 1965 by Daniel Weissbort and Ted Hughes and ran for forty-four issues. It was succeeded in 1986 by *Poetry World*, of which only two issues appeared. Beginning in 1992, *MPT* will again be appearing (twice a year), published now by King's College, London University. *Periplus*, like *MPT*, features European poets—a new generation of European poets—but at the same time explores an area hitherto neglected, the verse traditions of the Indian subcontinent. Almost a century into the age of translation, perhaps it is time to see if we cannot have a new balance of poetic power.

D.W.
A.K.M.

MARVIN BELL

Talk to the American Literary Translators' Association*

I hope it will come as no surprise to anyone in this room if I say that, to poets, you translators are heroes. But I may mean more than you realize.

The part you already know, I trust, is substantial enough. Certainly any writer is pleased if his or her work is translated into other languages. You know this. And certainly any poet should understand by now the difficulties of translating, including the rush of decisions that must be made by the translator and the serious implications of each. You know this too. Some poets may even fancy, for what seems good reason at the time, that whoever translates his or her work must surely love that work, and—who knows?—perhaps even its author. Well, a writer must have imagination!

But when I say that you translators are our heroes, I mean something more. I will tell you the story of it.

For twenty-five years, I have taught seminars to talented, skillful, individualistic, sometimes brilliant, poets—writers in their twenties and thirties, sometimes older—various in method

* The talk was delivered to the panel on translation and creative writing at the 1989 conference of the American Literary Translators' Association (ALTA), held at the University of Iowa, Iowa City.

and belief and, above all, independent and self-motivated. Whatever we may be reading together, we usually end up writing together as well. I have two rules about the assignments we make to ourselves. One is: Teacher has to do the assignment too. That's right: I don't make assignments I don't do myself. There is, in this, a lesson, I hope, that can be made only by example: writers are people who write. Writers are people you can't keep from writing. Indeed, writers are helpless to do otherwise.

'In the beginning', like they say, these were assignments of form. We might, for example, study sestinas one week, and the assignment, inescapably, would be to write either a sestina or a poem that sounded like a sestina but wasn't. So, too, with sonnets, villanelles, prose turned into poetry, emblems, prose poems, syllabics, accentuals, rhyming syllabics, rhyming accentuals, rhyming free verse, and so forth and so forth. There was the occasional agreement to surround a single subject.

Then came a seminar meeting—some years ago—in which we were discussing emotion and sentimentality in poetry. This time, the shyest person in our group—a student of Japanese poetry—contributed the assignment. She asked if we wouldn't agree to attempt to write poems which would be 'as emotional as they could be without being sentimental.' We agreed to do it, but in my heart I felt that such an assignment would fail. I wasn't even clear about the nature of the assignment. It would produce, I was sure, not poems, but sponges. Now, that would be ok. Our other rule is: no one has to write a *good* poem.

Still, because I was so certain that nothing worthy could come from such vague orders, it was the one time Teacher didn't do the work. Seven days later, however, when I plucked that week's new poems from my mail tray, I was amazed to find that they were indisputably terrific. They had a wholeness, a transformational character beyond the imagination of mere figures of speech. And they were not even sentimental! Indeed, I felt required, as I xeroxed their poems, to write my own. And I should tell you that the poem appeared in my next book of poems, and that it was then reprinted in my *Selected Poems*. In other words, I found that I would stand on the result.

Well, in that seminar we had been reading only poems in translation. Without, in general, judging the translations or comparing them with other versions. Without, in general, researching the lives of the poets. Without, in general, the benefit of a surround of commentary. In other words, we were reading these translations—even absorbing them—as primary works, poems, not at all as versions pointing to some other manifest ideal. Of course we tried to use both common sense and common knowledge.

To alter Valéry—surely every translator has said this many times in private—we might say that no translation is ever finished; it is merely abandoned. Valéry was referring to the fact that no poem could accomplish the abstract, ideal sense of it that existed in the poet's mind before that ideal was reduced to choices, forced to point this way or that, and generally turned into crummy, human language. The ideal, after all, is a complete form: a circle, if you prefer. Nothing gets in, and nothing gets out. Once, I suppose, we could have referred to the geometry of the ideal as a 'Berlin circle.' But recently, it is to be observed with joy, ideals seem to have gone to hell. (That's a joke.) In any case, the geometry of the ideal is nothing short of the geometry of the universe. Nothing gets in, and nothing gets out.

But the poet abandons his or her poem, it would seem, long before the translator abandons the source—or ideal—of his or her translation. Thus—and I was aware of this when I agreed to be a part of this gathering—some of you may find a poet's willingness to be influenced by a text without scholarly regard for its source and context to be contemptible. (Even though you are our heroes.)

Let me be clear that I am not speaking about 'imitations', but about influence. Let me tell you about the overwhelming importance of poetry in translation in the present lives of the poets of the United States and what sort of influence you have.

My role among you, I repeat, is that of the lucky ignorant whose message is that much work in translation seems even to the non-translator to have a weight and a range of imagination lacking in much poetry from our own country. My 'expertise' lies in the general area of ignorance, so far as translation goes. I have had a few ideas about certain translators here and there—you might say I have applied the Ben Belittmus Test when certain texts gave off a greenish light—but my ideas about translations have generally been like my ideas about wines: I may prefer the same bottle as does the cultivated crowd, but their explanation makes more sense. For what it's worth, here's my explanation.

In an interview many years ago, Robert Lowell was asked about American poetry and replied to the effect that American poets do a very difficult thing very well. Lowell may not have been thinking of technical matters, but indeed the manifest difficulty of American poetry is precisely what has made it so useful in the classroom. The New Criticism—I refer here to the *old* New Criticism—was a response to, and then, some would say, a cause of, the modernist model of the metaphysical poem: so visibly accomplished in its effects, alert to pun, nuance, ambiguity, conceit, myth, allusion, so dense, grave and woven; and, likewise, so civic, so responsible to a recognizable dramatic situation, a location, the poetic furniture of the moment, the apparent linearity of the outer life. If such a poem was a graph of the mind that created it, that graph could be drawn by putting the poem on the Understanding Machine. Even Williams, my beloved William Carlos Williams, had described a poem as 'a machine (large or small) made of words.'

It was a poem—it still is—good for conversation, full of puzzles and tricks, observing the self and the rest of the world with tilts and levels and swings and bellows. It was, above all, intellectually trustworthy. It proved its trustworthiness every few words by some act of attention to the very way in which something said earlier had been expressed. It was, to employ a common definition of the time—one that still applies to a certain kind of poem when it is good enough—it was like a spider's web in that, when any one part of it was touched, the whole of it trembled. Definitions of poetry at the time were usually descriptions of craft.

Then something began to change. Robert Bly's feisty magazine, *The Fifties*, was publishing more poetry in translation than not, and its editor argued, in an essay full of ridiculous hyperbole and manipulative metaphor, but still with a silo of truth at its center, that American poetry of the fifties was like a pinball machine full of effects. To make his case, he compared that poetry—rhetorical, fussy, grounded and, yes, very accomplished—to poems from other cultures—imagistic, surreal, abstract, employing the generalizing metaphor rather than the metaphor as a momentary figure of speech, emotional, visionary and, yes, often overtly political.

It took a long time for publishers to catch up with the translators but, by the mid-seventies, the translation and publication of contemporary poetry from other languages into English had blossomed. The poets themselves were so engaged. And the result was that one could see a new sense of poetry in the work of the poets of the United States.

Why? And why does it make such an important, and, I think, positive difference in the work of young poets to give them the example of poetry in translation? Is it only a matter of different styles?

No, I think not. It is, rather, a matter of *permission*. Whether from a lucky naiveté, or from the lifting of local restraints one feels when crossing certain borders, we have granted permission to poets from other countries that we have not granted to ourselves. The Russian poem praised for its expansiveness, its giant beating heart, when written by one of our own, is thought pretentious. The Spanish surrealist poem seen to contain a great original force, written by our poet, is called frivolous. The Chinese poem praised for its provincial simplicity and clear lines, if identified as a poem by a poet of our country, is dismissed as thin, simple-minded and stingy. The Hebrew poem admired for its profound tone of religious sorrow, identified as an American poem, is written off as church verse or the conventions of literary Judaism. The Greek poem respected for its historical scope and conversational intimacy, written by an American, is set aside as the work of an academic stylist. The subtle political allegory and irony in the poetry of any number of Eastern Europeans, if written instead by one of our poets, is ridiculed for using 'a device of the fifties' and 'playing it safe.' The plain, affecting speech of a poem in translation is seen to be a natural method. But so is the language of artifice, the language of dream and vision, the language of rapid association, the language of metaphorical philosophizing . . . In our own poetry the presence of any one of these may be considered merely a banal failure to do it elsewise.

A student told me that he wished he had a nickel for each line he had taken out of his poems because it did not seem sufficiently 'cool' to be part of an American poem.

We accord to poets from other countries a freedom of expression we do not permit ourselves. For the young poet, or even the perpetual beginner and the committed experimentalist—the three categories in which I happily place myself—permission and influence are inextricably linked. Thus, I have found that, when those poets in my seminars are set free to be influenced by poems so removed from their cultures that they are called 'translations', they are able to give themselves a permission beyond what their good-student lives usually allow. They learn to be influenced. Thus, they learn what it is to be original. The results are always striking.

Along the way, we confirm for ourselves the flip side of

Frost's famous remark that 'Poetry is what is lost in translation.' Mark Strand and Charles Simic, in their introduction to *Another Republic*, argue that poetry may be what is *retained* in translation. We may even, they assert, read poetry in translation—'oddly', they say—for *content*.

Maybe it's that sense of *content* that accounts for the greater permission we give poets in translation than we sometimes give ourselves. Wouldn't that be a kick in the pants of our current theorists and reviewers if it turned out that poems had content—not just theme and style?

Between what is lost in translation and what is retained in translation—one may argue that what is retained in translation is the *poetry* in the poem—between these stands the translation. Finally, it may enter the seminar room: loved and hated, found out or abandoned, just like a real poem, without much concern for how to distribute the credit and the blame.

I like the idea of having at least three definitions of poetry in mind at any time. One would be loosely based in craft (for example, 'Poetry is language that makes a difference'), the second would be based loosely in content (for example, 'Poetry is a lack of synonyms or, for today, 'Poetry is what is retained in translation'), and the third would be a combination (for example, 'Prose is prose because of what it includes, while poetry is poetry because of what it leaves out'). I don't think anyone can make a broad enough range of definitions of poetry from inside a single culture.

Well. The old metaphysical model that served so well so long can no longer keep up. The 'language poets', so-called, and the 'new formalists', so-called (the latter incorporate, in their promotional literature, those they believe to be poets of narrative—such would-be literary movements are in reality composed of Model-T lovers). Rather than create a poetry, as they are doing, to fit an old model of criticism, with its attendant dullness, we are far better off, I believe, in encouraging the internationalization of poetry to which you have contributed so much. Perhaps it will take us to a point where we will have a New New Criticism not obsessed with the inescapable relativity and subjectivity of language but, once again, with the descriptions of texts even unto ideas and emotion. The poetry that is retained in translation is, after all, a kind of emotional thinking. 'Nothing counts in a poem', said Pound, 'save the quality of emotion.'

Translators of poetry, and perhaps even those inexpert among us who take other poetries where we can find them—despite the irresolvable argument between those who, having to choose, choose to be faithful to the body, and those who, forced to make the same choice, prefer to be faithful to the spirit—despite all lingering academic questions about how to reproduce, those who translate poetry and those who absorb their efforts continue to make new poems. In other words, they make life.

Last month, I went to Wyoming. At the Snake River Institute, I taught a three-day seminar to a gathering of mostly unsophisticated poets. I sent a packet on ahead: I like to have some texts at hand when the talk turns to poetry. I used good poems—Bishop was in there and Wilbur and many others. At the bottom of the pile, I slipped in three translations, including Pablo Neruda's poem, 'Sexual Water.' I put Neruda at the bottom because, after all, such a poem could not be talked about as easily as the spider's web poems at the top. And besides, Americans tend to be practical, outward, impatient to get to the point. They like *USA Today* paraphrases.

So I put it at the bottom just to show the members of the seminar that there were other ways. On the second day, I asked them if they had taken a look at the handout and if there was a poem they wanted to talk about. Yes, there was. They all wanted to discuss 'Sexual Water.'

I can't do any better than that.

LUCAS MYERS

Translating Attila

Ágnes Vadas, my contranslator, saved up her pengoes and bought The Collected Poems of Attila József at the age of nine in Budapest. Before we were married, as a pastime, we translated a few Hungarian poems, working together. I had been attracted by the Penguin selection of poems by Sándor Weöres and Ferenc Juhász, especially David Wevill's translation of Juhász's 'The Boy Changed into a Stag at the Gate of Secrets', and she was pleased by my interest. We would sit in the living room of my house in the Catskills or in the yard looking out towards the opposite ridge, she would read a poem aloud and give me a literal translation, word for word, and I would try to produce an English version. One evening, she read József's 'Ode' and, since then, we have concentrated exclusively on his work. That was 1974. What had been a pastime became something more. We have now translated about forty-five poems. Virtually all that Hungarians consider 'essential' as well as a number of less celebrated ones are among them. They represent the phases of his mature work and his various styles and preoccupations.

József was born in 1905 to a woman from a peasant village in the central plain and a Transylvanian soapmaker who left her, their two girls, and Attila, the youngest child, when he was three years old. He lost his mother at fourteen, acquired, with the help of his brother-in-law, friends and patrons, a broad education in Hungary, Vienna and Paris, lived for five years with a committed communist in 'an alliance' which was 'not love', had a number of unhappy affairs with bourgeois women, subsisted in perpetual want, sometimes destitute and never securely employed, was overtaken by paranoia and schizophrenia, and threw himself under a freight train in 1937.

The acceleration of history and the passing of the old order between the wars opened fissures in Hungarian society even more dramatic than in most other European countries. In 1920 Hungary lost three fifths of its population and two thirds of its territory under the terms of the Trianon Treaty. The dominant political and intellectual themes of the period, Marxism, Freudianism and the rise of Nazi power, engaged József's powerful intellect. He was a chess player and also a masterful manipulator of Marxian dialectic and would sit up all night in cafés at one or the other pursuit. For a few years, he was a member of the illegal Communist Party; he brought food to imprisoned Party members including Mátyás Rákosi, who was the Hungarian Stalin after World War II until his deposition three months before the uprising of 1956. But the Party could not tolerate a member who did not believe in a dictatorship and was a sympathetic student of the writing of Sigmund Freud; it threw him out, though it claimed him as an ornament of Marxist literature some time after his death. Furthermore, his greatest poetry is infused with a numinous sense of the natural world and the cosmic order which wasn't 'scientific' enough to appeal to many of his Party comrades.

Arthur Koestler says that 'in Hungary, Attila József is posthumously regarded as the greatest poet the nation has produced.' We thought that a literal, even bald translation, as close to the original as we could make it, would bring into English not only something that didn't exist before, but something the like of which did not exist, the poetry being different from any I knew in my own language. Of course Hungarian and English words and syntax are not congruent, so being quite literal wasn't easy.

Furthermore, the verse of Attila (as most Hungarians call him and I, with the affection and spurious intimacy of the translator or biographer, will also call him), the rhythm, rhyme and metrical structure, is so remarkable and so much at the heart of the poetry that, in time, I came to understand that we could not bring his art across by semantic closeness alone. His verse forms are extremely varied. He wrote in most traditional forms, in free verse, and with folk prosodies. As often as not, intricate forms and rhyme schemes are of his own devising. Where Attila has a quatrain, I have four lines; where he has free verse, I have a parallel organization of the successive lines; where he has an invented stanza form, I follow that.

Hungarian is a Ural-Altaic language with the accent always on the first syllable. The feel of it is closer to Latin and classical Greek than to most modern European languages. It is unusually concentrated and unusually precise. Agglutinative, it says in one word what may require three, four or five English words to say. The sound of it is insistent, a succession of beats. Hungarian vowels all have both a short and a long form which gives the language rich resources of sound and, in their alternation, an unusual capacity for musical expression. It rhymes easily so that a translation of a heavily rhymed stanza into an English one with all the rhymed positions filled may sound like a barrel organ where the original does not.

My contranslator was not slow to point out cases where a literal version was too bold to create a response in the English reader strong enough to approximate the response of the reader of the original. Early on, I began to put greater emphasis on the English verse, still trying never to sacrifice the meaning and signification of the original. Since, in the nature of things, the English verse could not replicate the Hungarian without forcing unnatural diction or violating meaning, I sought forms which would create a similar response in English—by carrying a similar tone and mood—from the different music of the original. As for the rhyme, where Attila, for example, had abab, I generally put abcb. Besides avoiding a jingly effect from an original that didn't produce it, this practice had the advantage of enabling me to translate a number of poems that I was not able to translate into natural-sounding language with all of Attila's rhymes in place. In several ironic poems, I do duplicate the original rhyme scheme. A stanza from what we see as one of our successful translations, 'On My Birthday', reads as follows:

> 'As long as I have any say-so here
> you won't teach on this hemisphere,'
> he stuttered
> and spluttered.

The speaker here is Antal Horger, immortalized in the poem, who in 1925 had Attila expelled from the University of Szeged for publishing 'Pure in Heart', a dangerous poem in his traditionalist's view. But our translation, which duplicates Attila's verse form and is quite close to his meaning, is not as funny as the original. I think one reason is that the Hungarian sensibility is different from the English or American: the tone of self-mocking irony and sympathy with one's own dilemmas goes over better with the Hungarian reader than with the English, who might prefer more restraint. I don't know what to do about that.

Translating Attila was extraordinarily time-consuming except in the few cases where my version came in a burst. For long periods, sometimes years, we didn't look at the poems. When we did, we examined every word whose shades of meaning were delicate and I sat for hours (an infinite pleasure if there were world enough and time) looking for the right word. A single expression would be right, two others more or less acceptable, and five or ten further alternatives would ruin the stanza. I revised most of the translations quite a few times. Twenty, for instance.

In 1987, in a motel room in San Diego during a long business trip, I pulled out about ten translations which I hadn't worked on for several years, studied them with the originals, then without referring to either began each poem anew. A lot of the translations improved this way. Translations of new poems also began to get better. Something had happened. I suppose I had taken the poems into me, got some distance from them, and produced translations that carried my desiderata but also seemed to me more readable.

We never really developed a theory of translation, which in any case hadn't interested us very much, or even adopted a constant set of procedures. I came at each poem individually. I did have priorities. In the first place, be faithful. In the second, if a line doesn't sound natural in English, start again, while allowing yourself locutions that are uncommon (being derived from another language) but sound as though they could be natural even if uncommon—this being an enrichment. In the third place, try to develop a verse form which will elicit a response similar to the response the original elicits. In the fourth, rhyme as well as you can but do not allow yourself distortions or unnatural English for the sake of a rhyme and reduce the instance of rhyme if the barrel organ looms. The first of these priorities was not absolute. Occasionally, with reluctance, I abrogated it, generally in favour of the second or the third.

ATTILA JÓZSEF

Twelve Poems

Translated from the Hungarian
by Lucas and Agnes Vadas Myers

The Shoes

A poet's pulled me on his tiny
feet and trampled me,
that's why my two sweet broken cheeks
gaze out so mournfully.

The light might plank, the skies may smoke—
will the mud turn into jam?
He doesn't care, he pulls a lace
to cut wicks for his lamp.

I squish in pain, his face is stern,
he pays me no attention.
I might breathe out my soul and he
betray no apprehension.

Before his eyes I could spread forth
my heart upon the path.
He'd blush a bit, then poke at me
in a fit of wrath.

At times my tongue sticks out at him—
if he looks down to see

he gets so ugly that he will
stick out his tongue at me.

Once a hussar owned me and
the life I led was splendid,
though he kept changing faithlessly.
But now that life has ended.

He'd never deal with me this way,
dropped in the snow and lost,
left in the hoary North's thick gloom
to pulverize with frost

or in jungles of the sweating South
torn by a tiger's bite,
or home to lizards hearkening
to the rustlings of the night,

or hauled by termites in the dark
masticating me,
or lying on the shadowy sands
of the green depths of the sea

or in animal-flowers' silences
as a flower-shoe
on which the slithering flesh of snails
spreads a glaucous dew.

Oh, a macaque might smell me out
in the foliage of a tree
or hurl me down into a swamp
where a stork would pounce on me!

She'd lace me up with river snakes,
spear me and fly away,
or envelop me in feathers and
stand brooding where I lay.

I'm prisoner of the poet's feet,
  two years they've trampled me,
that's why my two sweet broken cheeks
  gaze out so mournfully.

1932

At the Edge of the City

At the edge of the city
where I live
the soot like tiny bats
through the crumbling dusk
flies on soft wings,
settles like guano,
thickens and grows hard.

The times lie on our souls that way.
Like thick rags of heavy rain,
they wash the scarred tin roof—
and pain wipes at the hardening stuff
in vain to get it from our hearts.

Not even blood could wash it off.
We're a new people, a different herd,
different in speech and in the way
our hair lies on our heads.
Not God, not even intelligence
but coal and iron and oil—

actual matter created us,
poured us out in the moulds
of this monstrous society,
fiery and hot,
to stand up for humanity
on the everlasting soil.

We follow bourgeois, soldiers, priests,
that's how we've become at last
faithful hearers of the law
and why the meaning of man's works
sounds in us,
like a viola's tone.

Destruction filled the past but never
since the solar system took its form
have so many caused to perish
what had been imperishable,
as arms, starvation, cholera
and blind faith raged in our homes.

Never under the stars have those
who would prevail in the end
been so humbled as we by you.
We looked down at the ground,
the secret which had been
enclosed there opened up.

The machine's a domesticated beast
but see how it's run wild.
Fragile hamlets crackle
like weakened ice on ponds.
When it leaps, the city's plaster
falls and the sky resounds.

Who's going to wave it down?
The shepherd's dog? Or landlords?
Its childhood is our childhood.
It was tame, raised up with us—
go give it its commands.
We can tell it what it is.

Yet we see it won't be long

till everybody kneels
and prays to something which
is merely his possession
but that will only come
to the man who gives it feed.

Here we are, matter's children,
together and mistrustful.
Lift up our hearts! The man
who lifts them up will own them.
Only he who's full
of us can be so strong.

Up with the heart, over the factories!
The man who's seen the sun
gutter in their smoke
and heard in the groins of earth
the pulse of its numberless tunnels
has seen and heard such a great sooty heart.

Up, up! around this divided land
the wooden paling cries.
It dizzies and staggers from our breath
as if a storm were going through—
Blow upon it! Lift your hearts!
Let it smoke up there

while our fine capacity,
the ordering faculty of the brain,
aware of the finite infinite,
renders clear the exterior
forces of production
and the instincts within man . . .

This song howls at the edge of the city.
The kinsman-poet watches,

constantly watches the soft fat soot
which falls, continually falls,
settles like guano,
thickens and grows hard.

The poet—words clatter on his lips—
yet he (the engineer
of the magic of the given world)
looks into the conscious future
and, as you will do externally,
within himself devises harmony.

1933

Ode

1

I sit here on a glistening
wall of rock.
The early summer breeze
drifts like the warmth
of a meal.
My heart gets accustomed to silence.
Which isn't so hard to do—
things which had faded reassemble here,
my head bends forward
and my hands hang down.

I look across
at the mountain's mane—the light of your forehead
shines from the leaves.
There's no-one, no-one on the road.
I see the wind
rustle your skirt,
and in the branches,
your tumbling hair,

your quivering breast
—from where the Szinva flows,
from its white stones
I see the laughter
burst from your teeth.

2

How enormous is my love
for you whose words distilled
loneliness, hatcher of plots
in the deepest cavities of the mind,
and the world's totality.
You separate from me
like sound from a waterfall
and gently run away.
Among the summits of my life
I reverberate from earth to sky
in the closeness of the distance:
I love you, sweet step-woman!

3

I love you as a child its mother,
as silent pits love their own depths,
I love you as concert halls love light,
the soul loves flame, the body rest,
I love you as living creatures love
to live until they die.

I hold each gesture, word or smile
as the earth receives what falls,
my instincts etch you in my mind
as caustic etches bronze,
your prized and lovely form,
the meaning of all thought.

The minutes clatter on

but you're silent in my ear,
stars explode and fall
but you're fixed within my eye.
The taste of you is cool
to my mouth like a silent cave.
Your hand on the water glass,
the delicate veins upon it
appear and reappear.

4

What substance am I made of
that your glance can carve and shape me?
What soul, what light, what power
enables me to wander
through the fog of nothingness
on the fecund hills
of your arching flesh?

Like the verb through the opened mind,
I descend into its mysteries.

The eternal current circles
like rosebushes in your blood
so love may bloom in your cheek
and your womb bear blessed fruit.
A net of tiny roots
weaves through your stomach's earth
in knots and imbricate strands
so the numberless cells of your fluids
and the tendrils of your lungs
breathe sybillate magnificence.

Eternal matter moves in tracts
of your intestinal tubes
and wastes attain rich life
in the boiling kidney wells.

Hills rise up,
constellations throb,
lakes move, works fabricate,
millions of living creatures swarm,
insects,
algae,
cruelty and kindness,
sun shines, aurora borealis glooms—
incognizant eternity
moves in that of which you're made.

5

Like clots of blood these words
are dropping down before you,
existence stammers and only
the law of nature is clear.
I'm reborn from day to day
by my diligent organs which
already prepare for silence.

Till then they cry your name—
you, whom they have chosen
from the many millions of women,
you, unique, sweet cradle,
sweet tomb and living bed,
take me into you.

(How high is this dawn sky!
Armies shine in its ore.
The great light dazzles my eyes.
I think I'm lost I hear
the clatter of my heart
beating overhead.)

6

(Side-Song)

(I'm on the train to follow you,
I may find you today,
you may cool off my feverish cheek
and maybe you will say:

The water's heated for your bath,
take this cloth and dry,
the meat is cooking on the hearth,
your bed is where I lie.)

1933

Awakening

1

The sky disbands from earth,
at dawn's pure word its flight
spins beetles and children out
into the early light.
No haze is in the air
and its glowing, weightless breeze.
Leaves like butterflies
have lighted on the trees.

2

I felt things rightly ordered
in the pictures that I dreamed,
dust filaments in place
blurred yellow, red and green.
Now the iron world's the order.
A dimness runs in my brains
by day moons rise in my blood
by night suns rise in my veins.

3

I'm thin and eat only bread.
Among nonsense-spouters I try
to find something more certain
than speckles on a die.
No beef's pressed to my teeth
no child's pressed to my heart.
The cat can't catch mice both
indoors and in the yard.

4

The world's stacked on itself
one thing pressed on another,
like a pile of kindling wood
it holds itself together.
That's how it's all determined.
Only what's not grows bushes.
Only what's coming has flowers.
What's now flies into pieces.

5

Like a slab of silence I stood
at the depot against a tree,
the grey weeds brushing my mouth
raw and oddly sweet.
Dead still, I watched the guard,
what he felt, and on the cars
his shadow leaping about
across the dewy coals.

6

Your suffering's within you.
Your explanation's without.
Your wound's the world, its fever
mounting is your soul.

As long as your heart rebels
you're prisoner but you're free
when you don't stack up a house
for a landlord to occupy.

7

I looked from under the night
at the cogwheel of the skies,
the glittering loom of the past
spun law from coincidence.
I looked to the skies once more
from under the haze of dream
and saw the web of the law
forever unravelling.

8

Silence listened—the bell
struck one. Your youth might be
relived; between dank walls
for a second you might feel free.
I got to my feet and saw
the Great Wains and the stars,
over the mute cell walls
their shining prison bars.

9

I heard the cry of iron
I heard the laughter of rain
I saw the past burst open,
abstraction alone forgotten.
I can love but can't unstoop
my burden's weightiness—
why do we have to forge knives
of you, gold consciousness?

10

There is no mother or father
in the heart of an adult
who knows his life is only
death's running supplement
and he can give it back
like something that he's found—
so he keeps it, he's no god
to himself or anyone.

11

I saw Happiness—she's blond
and weighs three hundred pounds,
her twisted smile was tottering
across the grass-cropped mounds.
She played in a lukewarm puddle
and grunted over towards me.
Among her downy bristles
the light stirred gropingly.

12

I live down by the tracks
trains come and go all night
their window frames hurl through
the fluttering dark and light.
That's how the light of days
hurls towards eternal dark.
Lit by each frame I stand
and silently watch their flight.

1934

At the Danube

1

I sat on the bottommost stone of the landing
and watched the rind of a melon go by.
Absorbed in my fate, I scarcely took note of
the murmur of surface, the silence of depths.
The river flowed through the core of my being,
the Danube was boundless and troubled and wise.

Each motion, each wave like the back of a workman
who's digging or hammering or patching a wall
would tighten and lurch then abruptly let go—
they were like my mother who cradled and rocked me
telling a story—they were like my mother
who took in the washing of all the town.

The rain was beginning to fall in a drizzle,
then stopped for a moment as though it were useless.
Like a man from the mouth of a cave I was staring
and saw the horizon through stretches of rain.
The past was indifferently, ceaselessly falling,
the prismed, eternal descent of the rain.

The Danube flowed on, its waves like a mother
absently dandling a child on her lap
played all around me,
splashing and gurgling,
gathering themselves on the current of time
and listed and fell like stones in a churchyard.

2

As though for thousands of years I'd been gazing
and abruptly the whole of time came clear,
all of my hundreds of thousands of forebears
were cognizant of it with me there.

I see what they didn't because they were hoeing,
coupling and killing as they had to be,
and I understand that engulfed in material
they saw the things I cannot see.

We know of each other as joy knows of sorrow.
The present's for them and the past is for me.
We write a poem—they're holding my pencil,
I remember their lives as they move in my veins.

3

My mother was Cuman, my father half Szekely
and half Romanian or maybe all.
The food was sweet from the mouth of my mother,
from my father's, the truth was beautiful.
Whenever I move, they are embracing
and sometimes this makes me sorrowful—
It's death. I'm made of it. 'You're going to see
If our life is over . . . ' They talk to me.

They talk to me since they're me already,
and though I'm weak, that's how I'm strong.
I comprehend that I'm my begetters,
my forebears back to the primary cell.
I'm the Forebear's mitosis into the many,
content to become my father and mother;
as they come apart and divide in two,
my soul's propagating into a Whole.

I am the world—what was and is:
All the peoples fighting each other,
past conquerors having their triumphs in me
and the conquered having their agony.
Árpád and Zalán, Werbóczi and Dózsa
Turks, Tartars, Toths and Romanians stir

in my heart, which owes its life to the past
and the peace of the future—for living Hungarians!

I want to work. It's struggle enough
finally to come to terms with the past.
The waves of the present, the past and the future
swell in the Danube and mix into one.
The battles our forebears fought in this country
will dissolve into peace in recollection.
We're charged with settling our common dissensions.
A task that's not a small thing to do.

1936

I Open the Door

I open the door. The vegetable smell
of cooled-off pots moves sluggishly
as though it were pushed. The claw-foot stove
across the kitchen is baring its teeth.

The room is empty. No one's there.
In sixteen years I've not forgotten
sitting on the oilcloth stool
and trying to keep from whimpering:

They've buried my mother—I understand.
But what I cannot understand
is that she is not. I can't be calm.
I should've been grown. (The sink stand shines.)

It's not important but I didn't touch her,
I didn't see my mother dead.
I didn't cry. I can't believe
that it will always be this way.

1935

The Guilt

I'm bad and full of guilt I guess
and yet I'm feeling fine.
I worry in this mess why I'm
not guilty since I am.

I'm guilty—there can be no doubt—
but every time I try
to name the crime it's something else,
some puerility.

I look for guilt the way a miser
looks for vanished gold,
I left a mother on its account
though my heart was never cold.

Some day I'll locate it among
the heroes of morality
and shrive myself by asking friends
to meet me over coffee.

And I'll confess I killed a man,
my father I shouldn't doubt—
in a clotted night I contemplated
his life's blood flowing out.

No lies: I stabbed him with a knife—
we're human after all.
And suddenly as though we're stabbed
each one of us shall fall.

Confessed, I'll wait (You'll have to wait)
to see who'll claim he's busy
and get away, who'll cogitate
or fall into a tizzy.

And I will notice someone who
warmly, with his eyes alone,

signals me: There're others here,
  your case is not unknown.

If my sin's childish and inane,
  be that as it may—
this world can grow to childhood
  and I will let it play . . .

If god exists—and I doubt he does—
  he needn't be my solution.
I'll get some help from living men
  and manage absolution.

1935

My Eyes Are Jumping

My eyes are jumping in and out.
If I go mad, don't injure me.
With strong arms gently hold me down;

when my entire being squints
don't show your fist since I won't see.
Don't snatch me back from nothingness.

Think that in this world I've got
no one and nothing and what I call I
I haven't got either. I chew its last bite

as I am finishing this poem . . .
Naked looks like lights in space
search in me. What wrong did I do

that as much as I signal nobody answers,
that she who's rightfully mine doesn't love me . . . Don't
believe in my meaningless guilt

till mould and rot have acquitted me.

1936

Loneliness

May insects march across your open eyes
and funguses sprout mould across your breasts!
Stare down the loneliness you send me to,
grind your teeth together, gnash your tongue!

I wish your face'd shift around like dunes,
your lovely face. When you need to stroke a man
because your vacant lap has disappeared,
I hope your fingers tangle up in weeds.

Look at this hideous wish—the wish is you.
If people silently surrounded you and stared
to see who could have rendered me so vile,
it wouldn't even make you turn a hair.

Who's that you're pawing? If you have a son
his greatest joy will be to twirl around
while you stand there and bat your eyes on him
and alligators bloat their sprawling guts.

Dead still, I lie stretched on my back in bed.
I see my eyes: you stare at me through them.
Die! I wish so wordlessly you'd die
I think I'm going to expire from it.

1936

*Greeting to Thomas Mann**

Like a child whose body is craving sleep,
who's already tucked in his silent bed
but begs you, 'Tell me a story—don't go'
(so night will fall less suddenly on him)
as his child's heart throbs and palpitates—
he may not know what he wants more
the story or just to have you there—

so we've asked you to come and tell us a story.
Speak as you used to—
we haven't forgotten.
Tell us you're with us, together here
because we're with you like all who're committed
to what is decent in mankind.
You know that the poet never speaks false.
Give us truth, not merely reality,
give us what fills the soul with light
for without each other we stand in darkness.
As Hans Castorp looked through Madame Chauchat
let us look through ourselves this evening.
No clamor will break on your measured words.
Say what is beautiful and what is evil
lifting our hearts to desire from mourning
for we've just buried poor Kosztolanyi.
The horror of more than one monstrous state
gnaws at mankind as cancer gnawed him
and shuddering we ask what still may appear,
what wolfish conceptions are coming upon us.
Will their fresh poisons course through our system?
How long will you have a lecture hall?
The thing is—speak and we'll keep our ardor,
the men still be men and the women still women,
free, full of grace, and all still human
though the times are diminishing human beings.
Please take your chair and begin your story.
We'll listen and some will merely gaze
for we are happy tonight to see here
a man of Europe among the herd.

1937

* Composed to introduce Thomas Mann at his lecture in Budapest in January 1937. The police prevented the poem from being read, but allowed the lecture.

You Brought the Stakes

You brought the stakes but not the flowers
you talked back to the world beyond
you told your mother you'd earn a fortune
but you sit around like a vagabond,

like a toadstool underneath a tree
(if anyone thought of you, that's how it'd be).
You're locked away in the Seven Towers—
don't think you'll ever walk out of there free.

Why'd you chew rocks with your baby teeth?
Why did you race just to fall behind?
Why didn't you do your dreaming at night?
Come on and explain what you had in mind.

You always kicked the covers off,
you always scratched at a sore till it bled.
You're clever if that's what you want them to think.
So when did the world begin, deadhead?

Who loved you? Who belonged to you?
Were you hiding out? Who wanted your life?
Well, bear up under it the best you can.
You've got no bread and you've got no knife.

In the Seven Towers you're locked away,
you're lucky to have a pillow of down,
you're lucky to have a fire to keep warm.
Now tuck yourself into sleepy town.

1937

*On My Birthday**

On my thirty-second anniversary
this poem arrives, a surprise for me—
pell mell
a begatelle.

I've surprised myself with a gift today
in a dark corner booth of this cafe
from myself
to myself.

Thirty-two years have floated away
and I've never reached subsistence pay.
That's the bounty
of my country.

I might have been a teacher then
instead of a grinder of the pen,
awfully sad
for the lad.

But in fact I never got to be
because the Master booted me,
a disaster
of a Master.

He suddenly issued me a tart
threat for my poem called 'Pure in Heart'
and society
was safe from me.

He guarded it from me with a sword.
My soul calls up his sacred words
and the ardent flame
of his name.

'As long as I have any say-so here
you won't teach on this hemisphere,'
        he sputtered
        and spluttered.

If Mr Antal Horger's pleased
I won't be teaching ABC's
        his delight
        will prove quite slight

since everyone who speaks our tongue—
not just the classrooms of the young—
        I'll reach
        and teach.

1937

* This translation first appeared in *Delos* (College Park, Maryland).

An interview with
JOSEPH BRODSKY*

'Esthetics is the Mother of Ethics'

GRZEGORZ MUSIAL: This is the magazine I'm working for, *Res Publica,* a monthly formerly published underground, but openly since 1986. It is the leading magazine for the younger generation of the independent Polish intelligentsia. Every issue is devoted to a special problem. This is the Prague Spring issue, this is the Viennese issue. Vienna is the symbol of the end of Europe, fin-de-siecle decadence. The question is whether we are faced with a new decadence of culture at the end of this century. So the question of spirituality arises, moral values. Of course there is no need for me to repeat something you know very well, but I would just like to emphasize that in the non-violent Polish revolution, the question of spirituality was a crucial one, because the Cross was placed in the middle of this revolution as a symbol of the readiness to forgive, the readiness to bear witness to the times, to suffer, to make sacrifices. So that is why I would like to ask the great poet about ....

BRODSKY: Quit the epithet, O.K.?

MUSIAL: O.K.

BRODSKY: So, what is your question, simply ...

* The interview took place in 1989, when Brodsky visited the University of Iowa to give a reading of his poetry.

MUSIAL: What is your attitude to religion? The same question you were asked a half-hour ago at the question and answer session at the University, and which you refused to answer. You even added: 'This is one of those questions one is not supposed to ask, nor to answer'.

BRODSKY: I, I don't think, well, for one thing, in my view, religion is a highly ecclesiastical persuasion, so to speak, is a highly personal, not only personal, an intimate matter, well, and you don't talk about your intimate concerns in public. That's to begin with. Secondly, to talk about them in public very often amounts to proselytizing, and one very easily slips into that mode, which may be good for society, but for the individual is almost invariably bad, even if you bring to the audience what you consider the ultimate truth, even if you are convinced that it will be better for them. I don't believe that one should proselytize. One should simply leave these things for people to decide on their own. Faith is a matter of arrival, well, people arrive to faith, they don't receive faith, in my view. Life generates that in them, and nothing can substitute for the efforts of life. It's actually the job that is best left, and best executed, obviously, by time. In part my reserve here is also the product of living in a country, which embraces or is embraced by several creeds. It's not a homogeneous country like Poland, and to, say, carry a cross in America, it may produce rather unpleasant echoes, say, in the mind of some, Ku Klux Klan. Secondly, some people who are of different religious persuasions may feel somewhat offended. This place is very much like Alexandria in the second or third century B.C., a marketplace of religions. Well, in a sense, it is all very well for the Poles to be Catholics or new Catholics. But in so far as you are within the confines of your community, it's fine. The moment you spill over the borders you find all sorts of other things, and a great number of conflicts, very bloody conflicts, in the human history, precisely were religious wars, so one should bear that in mind. Now, we can return obviously to the question of religion, but before I forget your initial point about the decadence of culture. Culture decays only for an individual. Talking about the decadence of culture is to promote a solipsistic view of reality. It's like ethics dying for a lecher. If you are a lecher, ethics is dead. If you are not and you are living next door to a lecher, your ethics are in full bloom. So I think it's a little bit premature and melodramatic to talk about the decadence of culture and, especially, the new decadence, et cetera. And I'd be terribly careful with all this. Indeed, what's happening in the culture is rather peculiar, well, culture is not an exception from other aspects of existence, though we all become now—it's my firm view—the victims of a new demographic situation in the world. There is a huge increase of people. There is an old educational structure, however, surviving intact. That is, the old educational structures in society haven't adapted themselves as yet to the new demographic realities. And for that reason, a great number of new arrivals are facing very old tenets. Speaking of culture, for instance, in the last, let's say, twenty or thirty years, in literature, very little qualitatively new has happened. However, the population of the globe has doubled in the last twenty years, so the net result of it is that the new arrivals, the new generations, are dealing, or are exposed to at best, material which is dated from their point of view. That is, the new generations don't seem to have produced something qualitatively new. So, therefore, the new generations are living off the old ties, the old cultural ties. Unquestionably, that creates either thirst or neglect on the part of the general population. In the West it's always a matter of distribution, because there is a great deal of wonderful words, even produced fairly recently, but they're not distributed. It's a capitalist country and you go by profits, and the man who tries to make a profit, as a rule, has a very limited version of what the market is like. He goes for certainties, not for the possibilities, and that's what limits him. Well, he simply does not assess the importance of these losses, and so on and so forth, and presumably very often he cannot. In the East . . . in the East, also for a variety of reasons, it has been regarded as prudent and useful to assert old values, which is quite all right, or course, and very often it serves as a kind of good for the intellectual, or cultural glue for the society, except that I don't think that this glue is terribly solid. I would think that the trouble—and now we're getting back to religion—the trouble with religion and with that sort of new Catholicism of yours is to remember where it was in the non-violence in Poland, et cetera, et cetera. The problem with those things—how shall I put it? is that in the final analysis they are ethical issues, and ethics, by itself, can't keep society together. You need something else, and I think ethics can be easily faked. There is nothing easier than to fake high principles. I think in order to make society indeed survivable, you ought to give society esthetics, because esthetics can't be assimilated. That is, a man has to become an esthetical being first. Esthetics, you see, in my view is the mother of ethics. And good though the Church may be about ethical matters, it can't produce art. To say the least, the treatment by art of ecclesiastical matters is very frequently far more interesting than the treatment of ecclesiastical issues by the Church itself. For instance, the version of afterlife given by Dante in his *Commedia Divina* is far more interesting than what you can find even in the New Testament, not to mention St. Augustine, not to mention other Fathers.

MUSIAL: I think the definition of the standards which the Church is trying to activate in Poland now relates rather to the question of morality than . . .

BRODSKY: Well, what you in the end have is somebody stressing morality, and what is the basis for that morality? Well, the basis for that is obviously the notion of God. And we've been living in a world which has been very busy negating the existence of any supreme being. Therefore, the people have to take morality or ethical principles on faith, which is fine and dandy, but it can be easily challenged. As I said, I think that esthetics is the mother of ethics, and the ethical principles emerge from the esthetics, and esthetics is tangible, it is more palpable, in a sense, than the subject of your faith.

MUSIAL: I guess that you refer more to the moral teaching of the ancient, pre-Christian philosophers like Marcus Aurelius, for example, and to others like . . .

BRODSKY: Let's not name names. What I'm trying to say is this, the human being makes his or her choice first on an esthetical rather than on an ethical basis. Take a child, take a baby, a one-year old who doesn't speak, whose experience of the world is zero. Well, mother holds up baby in her arms, the parents are having a party, and the baby smiles at somebody and cries at another. Well, in other words, he likes this person and doesn't like that person. That is, the child exercises an esthetical choice. Or his judgement is an esthetical judgement, see, but that is, esthetics do precede ethics, that's what I'm trying to say. And I think esthetics is a more sound basis to build a society, or civil society, upon, because in the end, you have to, when it comes to the moral choices—if it's based on the Church, on the faith, on the religion—then you in the end have to legislate, whereas esthetics make you a decent human being without legislation.

MUSIAL: Where do you place the metaphysical experience then, the spiritual experience of a single person, the knowledge which goes with that kind of experience?

BRODSKY: Well, I place it above the Church. The Church, or religion, is one of the many manifestations of the metaphysical potential of human beings. The reason I do this, and I have to get a little bit personal here . . . Well, I don't want to take too much of your time or of your page, but what happened to me was that my life went round such a course that I've read the Bhagavad-Gītā, the *Mahābhārata* the Upanishads, became acquainted with them before I got acquainted with the Old and New Testaments, and it's simply because the Bible wasn't available in Russia for my generation. I came to read the Bible only at the age of twenty-two or twenty-three. So, I read those things and they provided me with terrific metaphysical horizons, indeed, horizons. With Hinduism, let's say, to put it sort of vaguely, you get indeed a sense of tremendous spiritual Himalayas, one reach after another, et cetera, et cetera. Still it was clear to me that it wasn't my cup of tea. That is, simply, well, how shall I put it, biologically not mine, because though I could practice, and loved to practice, self-negation—not self-detachment, even, but self-negation—I practiced that for purely defensive reasons, because when you're grabbed by the police, et cetera, and beaten up—well, I think that's an extraordinary hell—they can't really get you, because you think you are not your body. Still, in the end, I felt it wasn't mine, and when I read the Old and New Testaments, and I read them one after another, well, I don't separate them the way it's done, for instance, in this country and the way it's presumably done in Poland too . . .

MUSIAL: No, not any more, it disrupts the unity . . .

BRODSKY: So, one thing I immediately sensed, that while the Old Testament is more congenial to me, partly because I'm a Jew—but it's not because I'm a Jew, it's simply a frame of mind, certain experiences—still I remembered the metaphysical horizons offered to me by Hinduism, and I realized that, metaphysically speaking, the Old Testament, not to mention the New Testament, is inferior to the metaphysical possibilities offered to you by Hinduism. This is why I always find myself in a difficult situation when somebody starts to talk about this or that particular church, because I think human metaphysical potential very seldom—well, that's my conviction—very seldom is fully exercised by this or that particular creed.

MUSIAL: On a certain level of religious experience one comes to a similar pattern of moral values. When you go through the deep experience of Catholicism, for example, you go through the same steps of metaphysical experience, exactly seven steps—a seven-step mountain Thomas Merton called it—as you do in Buddhism, right . . .

TOMISLAV LONGINOVIĆ—Eight in Buddhism.

MUSIAL—Eight in Buddhism.

BRODSKY—Good to see you gentlemen counting. But I'm more interested, well, how shall I put it, in matters of principle here. I spoke not so long ago about that to Milosz, and in literature you very often, a writer very often finds himself in sort of a double bind. That is, partly as a writer, as an active writer—well, writing, editing, correcting, correlating the things that he's saying to more or less the reality of society, et cetera, or trying to influence the society—he finds himself squarely in the Western tradition, in the tradition of the control of the will, in fact, a personal lyricism, imposing that upon the society or altering the society, whereas at the same time, simply by virtue of his occupation, he finds himself very much in the Eastern bind as well, in the self-negating position. So, one way or another, a writer is a cross between those two, and in fact, the value of literature, the value of poetry in particular, for a reader, for a general population, is precisely this, precisely that breech or that fusion of the self-negation and will. That's why literature is so, in a sense, attractive to us, or poetry in particular. So, I'm not interested in the seven types or eight types of ambiguity, because there are more than that.

MUSIAL—André Malraux said that the twenty-first century will be either spiritual or it will not be at all.

BRODSKY—It may be . . . Well, Malraux said so many things. The French are very fond of making up reasons, ever since La Rochefoucauld, and presumably before, I don't know. Milosz thinks that we are entering, the world is entering an entirely nihilistic stage. I am not so sure of that, although on the face of it, reality doesn't conform to any ethical standards, as we see it. It's getting rather paganistic. I think what may emerge—and this is one of my greatest apprehensions—what may emerge is a tremendous religious strife, not exactly religious, between the Moslem world and the world that is vaguely Christian. The latter won't be able to defend itself, the former will be terribly assertive. It's simply for the numerical reasons, for pure demographic reasons, that I perceive the possibility for such a strife. I am not a sage, I am not a prophet, I can't presume to say what the twenty-first century is going to be like. To say the least, I am not even interested in what the twenty-first century is going to be like. Well, I am not going to be there, for one thing, so why should I bother . . . And it was easier for Malraux, it was clear that he wouldn't be there, so it was easy to fantasize . . . The foreseeable future, that is, foreseeable by me, which again can be terribly erroneous, is precisely the conflict of the spirit of tolerance with the spirit of intolerance, and there are all sorts of attempts to resolve that conflict now. The pragmatists try to suggest that there is some equivalence between these principles. I don't believe that for a minute. I think that the Moslem notion of universal order should be squashed and put out of existence. We are, after all, six centuries older than the Moslems spiritually. So, I think we have a right to say what's right and what's wrong.

MUSIAL: I'd like to come back to the idea of my last question—I don't want to take too much of your time. On a more personal level, while listening yesterday to your poetry reading, I was very moved by the way you read. I felt it was more like a moan, more like a cry . . .

BRODSKY: It wasn't a moan, it wasn't a cry, it simply has to do with the prosodic nature of the Russian language.

MUSIAL: Which I understand, but with the exception of what you do with your reading.

BRODSKY—Well, maybe . . .

MUSIAL—Rhythmically, it's very . . .

BRODSKY—But it has to do with the meters presented, with the prosodic aspects of the poem, of this or that particular poem. But all literature that we have in Christendom—and Russian literature belongs in Christendom squarely, poetry especially—is a spinoff of the liturgical services, of the liturgical practices, of the hymns, if you will. So, there is this tremendous carryover of those, and poetry is an art of assertion, it's not an art of self-effacement. And the difference, of course, between the English delivery, delivery of that stuff in English, and the delivery of that stuff in Russian is simply a cultural difference now, because it's considered *mauvais-ton* in the English tradition to be assertive. It all kind of started about a century ago and so forth. *Bon-ton* means to be self-effacing. But I think it's a little bit ridiculous, because poetry is not an art of self-effacement. If you want to be self-effacing, you can take the next logical step and completely shut up.

MUSIAL: I see.

LONGINOVIĆ: I think one question which is connected to Russian religious philosophers at the end of the nineteenth and the beginning of the twentieth century—Shestov, Solovyov, Berdyayev, who thought that we were in a way entering the new Middle Ages with the Age of Communism—do you see any connection with the apparent dissolution of monolithic communism?

BRODSKY—Well, you see . . . how funny . . in a sense, they were right, because—well, Shestov never said anything of the sort, it was Berdyayev's idea—in a sense they were right, merely because looking into the future they couldn't see anything but darkness. But like all those prophesies, especially Berdyayev's prophesies, well, it's basically easy to make a prophesy because we have so few options in general. Now it's seventy years later and I don't think it's going to be the Middle Ages. I don't want to sound overly optimistic, but communism, well, it's curtains for communism, for any sort of ideological society, in my view, at least in Eastern Europe. In the Orient, perhaps, and in Latin American countries, for perhaps the next thirty or forty years, and perhaps in the Far East, there is some possibility of some ideological society. But for the Europeans, it's over, so I think the societies now are going to evolve along a more pragmatical line, which is also nothing to cheer about. But one thing about a pragmatical principle is that it considers all the options and, therefore, it doesn't rule out the religious or metaphysical option, or won't legislate against those options. In that respect, we should not forget about one thing. We shouldn't say that the religious man is better than an atheist, though perhaps in our hearts and in our minds he is, and it's a more interesting, infinitely more interesting existence. But I think the root of all evil is when one person says, 'I'm better than the other'.

MUSIAL: You mentioned the importance of Milosz in your life, the influence he had on you. Could you say something more about this? I guess it's a kind of spiritual experience of the sort we were talking about before.

BRODSKY: Milosz is simply, well, a tremendous presence. I am fortunate to know him personally. First of all, I've translated some of his poems. Secondly, he helped me enormously. He wrote me one letter, a very short letter exactly at the moment of my arrival in the United States, which indeed took instant care of a great many insecurities that I harbored at that time.

He said in that letter, among other things—he was talking about translation, et cetera, et cetera—he said I understand that now you are worried about being capable of continuing to write in a foreign country, et cetera. He said, if you stop, if you fail, there's nothing wrong about that. I've seen that happening to people. It's perfectly human, et cetera. Well, it's perfectly normal for a human being to be able only to write within his own walls, or in his context. However, he said, should that happen, it will show your real value, that you are good on the domestic. There's nothing wrong with that. But when you read that sort of thing—no! And for that reason, I'm awfully grateful to him. But aside from that, I am envious, well, I have a terrible envy of him having lived such a long life. I wish I were there in the twenties and the thirties. I wish I had the same experiences. But ultimately I simply admire his mind. It's a Manichean bent. We make war about this and that, but there is nothing better than an argument with Milosz, it enriches you enormously. It's not only cultural differences or cultural material, baggage that he carries, it's the methodology of his mind, the entire predictability of it. It's great to talk with him about esthetical matters, about, for instance, matters of absurdity, of the absurd in literature, better than to talk about ethics, which is a much more stale discourse in the end.

MUSIAL: Have you ever had another deep experience with Polish literature or a Polish writer?

BRODSKY—Norwid.

MUSIAL—Norwid, I see. And contemporary writers, Gombrowicz, for example, who's had a tremendous influence on the young Polish intelligentsia . . .

BRODSKY: That's understandable, but I think for me, Gombrowicz, much though I like *Ferdydurke,* and what's the other one? I forgot.

MUSIAL— *Cosmos, Pornographia.*

BRODSKY—Much though I like those, I still think Gombrowicz, by the virtue of circumstance—not virtue, by the vice of circumstance—has come to make a big deal out of himself, for himself. That is, he was indeed a literary person, that is, temperamentally, he is quite the opposite, well, of me personally, because I don't write that much. I don't make literature out of everything. I used to, not exactly pride myself on it, but I used to regard myself more as a gentleman who occasionally occupies himself with composing a poem, or writing a piece, whereas Gombrowicz was indeed, well, his life was literature, and also he took himself and the unpleasantries [sic] of his existence a bit too seriously, in my view. But that's my view, that in no sense detracts from Gombrowicz. It simply my personal view of him. Well, now, for other Polish writers, I think the greatest boost I ever got in my life was from Norwid. I translated Norwid into Russian. Not much, about six or seven lengthy poems, and I don't think there is any greater poem in any language that I know of than 'Bema Pamięci Raport Załobny' [A Mournful Report in General Bem's Memory]. I happen to know that by heart, too, but that's one. It is simply the vector of tragedy in his voice. To me, he is a more important poet than Baudelaire, of the same period, a far greater thing, though I don't really care for his long dramatic poems. But the poems, some of them, are absolutely remarkable, terribly far ahead of his time. I don't place a great deal of value on that sort of thing, but it's stunning to find this sensibility in the nineteenth century. You know, in Rome on the via Sistina there are two houses, they are door to door, and there are two memorial plaques. On one it says here from such and such—and it's the same years practically, well, with one or two years' difference, they're almost overlapping—here lived Norwid, and the next door, here lived Nikolai Gogol.

MUSIAL: What do you think of Herbert's poetry?

BRODSKY: Well, I translated Herbert. I like Zbigniew enormously, and you are tremendously fortunate in Poland to have in one century, or rather in one half-century, poets of this magnitude, Milosz and Herbert. I would throw in for good measure also Szymborska, and not Szymborska entirely, but she has tremendous poetry, tremendous poems. I wish I were assigned to make a selection of her poems. But speaking of Herbert and Milosz, I can't really say one is better than the other, et cetera, because at those heights there is no hierarchy. However, in a sense, because Milosz' operation is huge, you can make the argument that Milosz includes Herbert's idiom, whereas Herbert doesn't include Milosz'. But there is no hierarchy at those heights, it's not even a qualifier here. I would say Milosz is a greater metaphysical event for me than Herbert, although I wouldn't like to live solely with Milosz' poetry and without Herbert's. It in fact makes me appreciate Milosz more and the presence of Milosz makes me appreciate Herbert more. He is in the final analysis, Zbigniew, a great esthetician, a great esthete. He is exactly the man who made his preferences, his choices, on the basis of taste, not on the basis of morality, on this and that. He said something to that effect himself.

MUSIAL: He wrote a poem, the famous one, 'The Power of Taste'.

BRODSKY: Yes, 'The Power of Taste'. I don't follow Polish poetry right now very closely, and I don't see now . . . Well, I used to, but for the last ten or fifteen years I've been somewhat out of touch . . . I don't really know who in general is what. I don't know who are the followers of Milosz, let's say. But at least I know who is the follower of Herbert. Zagajewski seems to me to represent a development of Herbert's idiom in many ways. He's an absolutely remarkable poet, and the most important discovery for me in Polish poetry for the last ten or fifteen years, and in fact, we're friends with Adam, and it's one of the better friendships that I've had in my life. That's all.

MUSIAL: Thank you.

TOSHIKO FUJIOKA

Five Poems

Translated from the Japanese by
Kazua Honda, Jon Silkin and the author

Knocking on the Door and . . .

Far off, the bamboos appear in the moon's brilliance; fresh
bamboos sprouting, and budding, in the sky's sloping line
to spread their leaves in the neighbourhood of the moon.

They stretch roots through to the world, beneath, and space
becomes their heaven. Their roots hang silver threads
through the sky

and thin silver strings root in the dark earth. In the night
moon, as they grow fresh roots leaves rise softly and calmly.
Waves brim the shore of the moon.

In which bamboo joint, I wonder, has a beautiful
Kaguyahime,* a graceful woman, travelled with her soul
full?

* Kaguyahime was born (as a baby) in a bamboo plant, and therefore her grandparents brought her up. She grew up to be a beautiful girl and was named 'Kaguyahime' because of her beauty, which shone. One day she said, 'I must return to the moon. I don't care to be a human being in this world. I belong to the moon.' And on the night of a full moon she returned to it.

This well-loved story is amongst the oldest in Japan and is referred to as 'the parent of stories'.

I pray that I can live such a life, such existence on earth. I see a candle burning.

Distantly in the moon the bamboos show. In the depths of the bamboo-stems new buds burn, and emit ceaseless light. The buds knock upon the gates of the future, which supports the slanted sky helping the growing buds to a full vigour.

Blossom

I ask you, take the light ash
of my bones
and bury this handful in your country
I cannot come to.

I am sending this bone, reduced
by a flame of aurola. I am white sand.

How beautiful the castle walls you have
which I am acquainted with
only by your postcards; in your hands,
please, carry my ash within the wall
and as if you planted seed, bury it.

My bones will be dead and asleep in earth
when you have seeded me. They will sprout
some day;

the trees growing by your castle wall will flower
a pink light wind, my nation's flower, Sakura,
cherry blossom. That flower.

Now my heart wakes
in your country
and starts to blossom
without hesitating.

Oh my love, live long
and distribute my ashes.

My child, my distant child
return my white sand,
my white bone
to the town circled by ancient walls.

A Room

After shutting the door behind me, I go for a while and discover a second door, and opening this into another room find a further entrance. Many doors that lead to one mysterious room, where many blossoms ripen and flower in fragrance.

The sun smiles through glass above the heads of Ling blossom.

With softness, in sadness, a Rape blossom says to herself, 'I must leave tomorrow,' and the others are affected by what she whispers, their corollas leaning against each other.

Many voices whisper; 'I have to leave, too: I must,' each blossom of Rape says to the other, with eyes intent on the eldest among them. 'Yes,' the older flower nods, and says, 'it is our obligation to supply more time to a candle's life when its tallow runs short,' and, 'we may add life to a candle that is about to sink into the dark.'

The candle shines a solemnity in the deepest spot of the flower. With tenderness the sun smiles.

Mellow Ripe Wine

An angel leaves heaven and by night enters the tightly-closed wine cellar. She knocks soundlessly on the door; and the jar stationed in the chill dark cellar has longed for such a visit as she, the angel, promised to make long past.

The cellar is now light as day, the jar silhouetted against the open square of time in which it has breathed for years. The jar inlaid with rich, coloured pictures and sculpture is lidded. Dry mud seals its mouth, cord securing its stopper. It faces its future, does not loose itself for that, or allow its lid to be removed.

Before time matures, what is in the jar is not visible. With pain, the angel blesses its capacity. Her eyes are bright, and the jar shines. A sweet smell and mellow sweetness must have existed in the contents rejecting all save the angel.

On this marriage night, the amber of the wine will make its journey formed by the stars.

Yukitsuri

A sound of water moving about a castle town, which I grasp as I trace my way along the earth walls. The sound is faint and like a spring. I listen to the hidden falling water leaking from icicles in the eaves, my heart filling with serenity: The light is so bright. The thin stream in front of the house runs with clear water in which melting snow dissolves. Loitering in a narrow street, I catch sight of a cap-shaped pine tree.

'YUKITSURI'*

Such a tender naming!

I see a cone negotiating the intense moment of a rope which holds branches and the earth. The pine tree supported by the earth has its specific points on the circle which the branches and the knotted rope make. The ropes and their lined-up knots point to the void, and the taut strings of rope from the branches show how tense they are. The ropes pull their branches away from the void, ah, no, support them with difficulty. Don't slacken. Don't relax. You have to bear the snowy weight. You have to bear the depth of the dark. The ropes strain now with the weight of snow in the branches. The ropes aware of the anxious heart. Many ropes in the void. They will be a lyre in night's white snows.

* Yukitsuri is a device to prevent trees from breaking in heavy snow. A pole is put next to the tree and from the top of the pole a number of ropes are tied to the tree's branches. The assembled ropes look like a cone.

A. K. RAMANUJAN

Lalla's Sentences

Reworked from materials in *Lallā Vākyāni*
by George Grierson and Lionel D. Burnett,
London 1920

Lalla (also known as Lalla Ded, Lalleswari, etc), the great woman-saint and poet of fourteenth-century Kashmir, is a byword in the country. In her, Sufi and Hindu (Saiva) mysticism found piercingly beautiful poetic expression. Her *Vākyāni* ('sayings, words, sentences') are on the lips of both Hindus and Muslims in Kashmir to this day; many of her sayings are proverbs in daily use.

The Grierson-Burnett book cited above contains the Kashmiri text of Lalla's poems, Bhaskara's Sanskrit translations, Grierson's English prose renderings, notes, grammar, and vocabulary. In 1914, Grierson's friend, Pandit Mukunda Rama Sastri recorded the text from dictation from an old brahmin in a village, added his Hindi and Sanskrit commentaries, and sent the manuscript to Grierson.

Readers who wish to know more about Lalla, her poetry, religion, and the legends of her life may consult Grierson and Burnett, Richard Carnac Temple's *The Words of Lalla, the Prophetess* (Cambridge, The University Press, 1924), and J.L. Kaul's *Lal Ded* (New Delhi, Sahitya Akademi, 1973).

The numbers refer to the Grierson-Burnett text.

8

Siva. Kesava. Jina.
Or that other one born in the lotus.

Let him bear any name whatever,
only let him cure me,
sick woman that I am,

of this disease,
this world,

whether he is this he,
that he, or that other he.

9

When the sun was gone
the moon came
with that round face of his.

When the moon was gone,
the mind remained.

When the mind was gone,
where did they all go,
all things born of sight,

earth, ether, sky?

10

Get up, woman, and make your offering,
the meat and the wine
in your hand.

If you know the Syllable
that is itself
the supreme place,

it matters little
what rule you break.

What's there to lose?

12

Take the kingdom and rule it.
You find no peace.

Give it up.
You still find no peace.

But the self free from greed
has no death.

And if, when it's alive,
it should die,

then that's true knowledge.

13

O Siva, god of the dark blue throat,
you have your six powers*

and I have my six.**

And yet, estranged,
I've fallen into grief.

Though between you and me
there was no difference,

there was still one:

you were the lord of the six
and I, by the six, was led astray.

* An enlightened ascetic's six qualities: omniscience, contentment, knowledge of the past (lives), sufficiency, potency, omnipotence.

** Often interpreted as sexual desire, anger, avarice, arrogance, delusion, jealousy.

17

The idol in the temple is a piece of stone.
The temple is a pile of rocks.
From crown to foot, it's all one stuff.
O learned pundit, what's this you worship?
Just bring to it a mind and the breath of your life.

18

Let him abuse me a thousand times.

If, by nature, I'm for Siva,
my heart will not grieve.

When was a mirror fouled
by a scatter of ash?*

19

Weary, they came
and came again.

Then, weary,
they had to go.

Forever, night and day,
they must move on.

Wherever they came from,
back there they must go.

What are they? What's anything?
Nothing, nothing,

and more nothing.

20

Even if you know everything,
be as a fool.

* Ash was used to polish metal mirrors.

Even if you can see,
be one-eyed.

In all things, be a thing
without sense.

That's the practice
that will bring you

to What Is.

24

Worth. Reputation.

They are water
        carried in a basket.

If a man is strong enough
        to grasp the wind in his fist

or tether an elephant
        with a hair on his head,

if he can do such feats,
        then he can also hold on

to worth and reputation.

25

By mastering the vital breaths
I cut through the six forests
till the moon woke up in me,
digit by digit,
and the world dried up in me.

With love's fire,
I parched my heart
as one parches grain

and in that moment
reached Siva.

26

My thought-steed speeds over my heart.
In the twinkling of an eye
he travels
a hundred thousand leagues.

If one does not know
how to seize the horse by the bridle,
the wheels of his chariot
will explode and fly into pieces.

28

Wear only enough to ward off the cold.
Eat only enough to quell your hunger.

O mind, give yourself
to seeing self and the Supreme,
and recognize your body
as nothing more than food
for forest crows.

38

Stop a flowing stream.
Cool a raging fire.
Walk the air with your own two feet.
Milk a wooden cow.
All these are, in the end,
but tricks of jugglery.

42

You alone are the heavens.
You are the earth.

You are the day, the air, the night.
You alone are the food I offer,
the sandal I rub,
the water I sprinkle.
You are all that is.

What then can I offer you?

46

It is He who laughs,
who sneezes, who coughs, who yawns.

It is He who bathes,
over and over, in holy pools.

It is He who is the ascetic,
naked from year's end to year's end.

Recognize Him. He is near.

47

It's a lake so tiny
a mustard seed can find room in it.

And, into it, deer, jackals,
rhinos and sea-elephants

keep falling, falling,

almost before they have time enough
to be born.

48

I, Lalla, tired myself out
seeking and searching for him.

I labored and strove
beyond my strength.

I looked for him
till I saw the bolts on his door.

Then, even as I was,
my love for him rooted

right there, where I was,
I gazed on him.

49

I burned my filth in my self.
I slew my heart.
And then, Lalla, my name,
spread abroad,
even as I sat, just there,
with bended knees.

50

I remember a lake
overflowing three times.

I remember once seeing the sky
as the only place there is,

and once seeing a bridge,
north to south, from Haramukh to Kausar.

And seven times, I remember, the whole world
becoming Nothing.

99

Heedless fool, lift your foot at once
and begin your journey.

It's dawn,
seek now your friend.

Make yourself wings
    and rise now
    on winged feet.

100

Give breath to the bellows
    as the blacksmith does.

Then your iron
    will turn into gold.

It is dawn.
    Seek now your friend.

101

I locked the doors and windows of my body.

I seized the thief of onions, my breaths,
    and called for help.

I bound him fast in the closet of my heart
    and, with the whip of OM
        I flayed him.

102

I, Lalla, went forth
in the hope of blooming

like a cotton flower.

The cleaner and the carder
kicked me hard.

The spinning woman lifted me
from the wheel,

all gossamer,
all made from me,

and then in the weaver's work-room
they struck me

as they hung me.

103

When the washerman dashed me
on the washing stone,
he scrubbed me hard
with fuller's earth.

When the tailor worked me over
        with scissors, piece by piece,
then I, Lalla, reached my end.

104

I didn't put my hope in it
        even for a moment.

I didn't trust it even by a hair.
Still I, Lalla, drank the wine
        of my own sayings.

Yet I seized that inner dark,
        brought it down,
tore it and cut it to pieces.

105

At the end of moonlight
I called to the mad one
to soothe his pain with love of god,
        crying aloud, 'It's me, Lalla!
        It's me, Lalla!'
and I awakened my lover.

I was one with him then and my mind
lost all its filth
in that lake of crystal.

106

With a rope of untwisted thread
I tow a boat on the ocean.
Where will my god hear me?
        Will he bear me across?

Like water in goblets
of unbaked clay,
I waste away.

My soul is dizzy, all in a whirl.
        I want to be home.

107

The sling of candy load
has come loose on my shoulder.
My day's work has turned crooked.
        How shall I succeed?

My teacher's words fall on me
like blisters, like losses.
My flock has lost its shepherd.
        How shall I succeed?

K.Pr. 57

One minute, I saw a stream flowing.
The next minute, I saw no bridge,
nor any way of crossing the water.

At one time I saw a bush blooming.
At another, I saw neither a flower
nor a thorn.

One moment I saw Kunti,
mother of the five Pandava brothers.*
At another, I saw only
the aunt of a potter's wife.

K.Pr. 57

Father,

take away from me
the longing for the unattainable.

Release the pigeon
from the pigeon-hole
of my heart.

My arms are tired from making
other people's houses.

O body,

when will they carry you forth
from your house?

Afterwards, a thousand people will come
waving their arms.

They will come and set you in a field
and lay you to sleep on your right side.

* The Pandava brothers are the warrior-heroes of the epic Mahabharata. In exile, they once live with a potter's family.

ARUN KOLATKAR

Nine Poems

Translated from the Marathi by the author

Pictures from a Marathi Alphabet Chart

Pineapple. Mother. Pants. Lemon.
Mortar. Sugarcane. Ram.
How secure they all look
each ensconsed in its own separate square.

Mango. Anvil. Cup. Ganapati. Cart. House.
Medicine Bottle. Man Touching his Toes.
All very comfortable,
they all know exactly where they belong.

Spoon. Umbrella. Ship. Frock.
Watermelon. Rubberstamp. Box. Cloud. Arrow.
Each one of them seems to have found
its own special niche, a sinecure.

Sword. Inkwell. Tombstone. Longbow. Watertap.
Kite. Jackfruit. Brahmin. Duck. Maize.
Their job is just to go on being themselves
and their appointment is for life.

Yajnya. Chariot. Garlic. Ostrich.
Hexagon. Rabbit. Deer. Lotus. Archer.
No, you don't have to worry.
There's going to be no trouble in this peaceable kingdom.

The mother will not pound the baby with a pestle.
The brahmin will not fry the duck in garlic.
That ship
will not crash against the watermelon.

If the ostrich won't eat the child's frock,
the archer won't shoot an arrow in Ganapati's stomach.
And as long as the ram resists the impulse
of butting him from behind

what possible reason
could the Man-Touching-his-Toes have
to smash the cup
on the tombstone?

Biograph

Knotting the cord, the midwife said,
It's a boy, it's a boy, it's a boy.
Piercing an earlobe, the goldsmith said,
Two bucks, just two bucks.
Syringe in hand, the nurse said,
It's not gonna hurt, not a bit.

Measuring my dick, Baban said,
Mine's bigger, bigger than yours.
Punching my back, Baban said,
My dad can lick your dad.
Kicking my shin, Baban said,
Sissy, a sissy, what a sissy you are.

Pressing her toes against mine, Bunny said,
Bicycle, bicycle, let's play bicycle.
Rubbing spittle on my tummy, Bunny said,
Doctor, doctor, let's play doctor.
Tickling my ribs, Bunny said,
Come on in, between the sheets.

Boxing my ears, a teacher said,
How much is thirty three times thirty eight?
Rapping my knuckles, a teacher said,
And where's Sheffield then? Where's Sheffield?
Squeezing my thigh, a teacher said,
Let's go to the mango grove.

Twisting my neck, the barber said,
Don't move now, don't move.
Measuring my chest, the tailor said,
Thirty one inches, just thirty one.
Forcing my foot into the shoe, the cobbler said,
Use it, and it won't be so tight.

Jumping on my back, junior said,
Giddyup, giddyup.
Giving me the boot, my boss said,
I can't help it Mr. Nene, I just can't.
Grabbing my cock, my wife said,
I'll chop it off one day, just chop it off.

Feeling my balls, a doctor said,
Hydrocele, I'm sure it's hydrocele.
Sticking a pin in my toe, another said,
Leprosy, you can take it from me, it's leprosy.
Tapping my stomach, a third one said,
Ulcer, ulcer, no doubt about it.

Stepping on my toes, a guy said,
Sorry man, I'm sorry.
Sticking an umbrella in my eye, another said,
I hope you aren't hurt.
Bearing down on me, full tilt, a trucker said,
Can't you see where you going you motherfucker?

Crabs

Look, look.
Just look at them.
The crabs.
There are two of them.

They're keeping watch.
On whom, you ask?
On you of course,
who else?

See how they're looking?
Looking at you,
naturally.
And you'll never catch them blink either.

One on this side.
One on the other.
At an angle of a hundred and sixty degrees
to your left and to your right.

They're going to eat your eyes.
That scare you?
It needn't, you know.
It's not as if they're going to start eating right away.

No. But one of these days.
Tomorrow? Who knows?
If not tomorrow, then the day after.
Or ten years from now, who can tell?

They're in no hurry.
They have plenty of time.
And they can live without food
for a long time, you know.

Look this way,

quick.
Don't turn your head.
Just move your eyeballs.

Do you see a crab there?
Not the whole crab, may be,
not yet,
but you did see something move?

Now look the other way.
No, no. Not the whole head.
Just move your eyeballs
like I said.

All you can see for now
is just the pincers may be,
but you'll see,
you'll see the whole crab yet.

And you'll see it clearly.
They're only doing their job of course,
but patience
is one thing you should learn from them.

The crabs belong to you,
and to you alone.
They have no interest in eating
somebody else's eyes.

They came out of your head.
Where else did you think they came from?
But how they've grown.
Look at them now,

big fat crabs.
They've been playing a waiting game
ever since they emerged
from your head.

They'll come for your eyes
any time you say.
Sometimes I think they're just waiting
for your permission.

All you have to do is give the word.
And once they've eaten your eyes,
their job is done.
You'll never see them again.

The Wind Song

Fuck your cap.
If it's gone, it's gone.
You still got your head on, right?
Hold on to it tight.
And look after yourself, man.
A dangerwind is blowing.
Look at all this rubbish in my hair
and damn this dust,
it's all getting in my eyes!

A window shuts with a bang
and showers glass on a fat mattress.
Chandeliers have too many problems of their own
to notice how the carpets have begun
to roll up by themselves.
A Ming vase, overturned,
spins around and rolls from side to side
on a newly waxed floor
as it wrestles with a private demon of its own.

A nylon sari slithers off the clothesline
and goes swirling up in the air,
pursued by the pyjamas
of a Bengali chartered accountant

that tear themselves away from a balcony railing
and, pedalling furiously, ascend to heaven
to catch up with the sari
and conduct a series of joint manoeuvres
high above the city skyline.

Like illiterate teleprinters
invented by a race of giants
before they could discover electricity,
rooftops come to life;
and the news of the coming storm is relayed from roof to roof
as tiles start talking in tongues.
Let me help you sweep the floor, Mr Kulkarni;
your college degrees,
they're all lying shattered on the ground.

Larry, Larry my boy,
has your dad been missing from his silver picture frame?
I thought I saw him running around in circles
around a fire hydrant in Apollo Street.
Last I saw him he was about to get into a fight
over a cute little piece of tinfoil
with three almond leaves,
two bus tickets and yards of carbon paper
that had ganged up on him.

Your poem, professor;
you say it's gone?
Slipped out of your hands and disappeared?
Just like that?
May be it went for a walk, you know.
Have you looked everywhere?
And don't look so sad, professor, cheer up!
May be the poem was never any good to begin with!
Have you considered that?

Drop the paint brush.
Leave your cans of paint and run, painter, run!
I hope you're fast on your feet.
The long legged foxy lady you painted on the billboard
has begun to shake her hips;
and I hope you're far away when she starts looking for you,
slicing her way through everything
that lies in her flight path
like a guillotine unhinged.

Teacher, teacher, look, the map!
The map of India!
The way it's kicking,
the way it's dancing on the classroom wall.
Do you think it's going to fall?
There it goes, it's gone already.
Cities, mountains, forests, rivers and all,
gone,
out the window and up into the sky.

To a Cloud

I will have to describe you as an explosion of cheeks.
I'm impressed.
I cannot think of anyone else who offers such a wide range.
In all sizes and in all possible and impossible colours.
I'm not being sarcastic.
Mind? No. Why should I mind?
Two in the front and two behind
is more common,
but you can have as many cheeks as you want.
There is no rule that says you can't.

I don't know who it belongs to—
which animal, real or imaginary,

and it could be anywhere in that jumble of cheeks—
but I'm sure there is an eye there somewhere.
And a nose too.
It's somewhere around.
Not necessarily on a face,
but then I do not know of any rule that says
that a nose always has to be in one place.

And even if there were such rules,
I'm sure they won't apply to you.
Because you're a special case.
For you, Sir Cloud,
everything is allowed.
For you, above all, are you.

You come rolling in through that window,
wiggling your arse
and smiling from cheek to cheek to cheek to cheek,
picking your nose that's sticking out of your crotch,
scratching your balls, three of them, hanging between
    your shoulderblades,
and sticking your tongue out of that arsehole in your armpit.
You come rolling in through my window
as if this room belonged to you
and you're welcome of course,
you can stay for as long as you like.
Make yourself comfortable.
But I don't have to tell you that,
do I?

All I'm saying is, on the other hand,
if you want to get the hell out,
IF I'm saying, mind you, IF you want to get the hell out,
get out of my sight and leave me alone
to study the ceiling in peace

for the rest of what remains of this night,
that's all right too.
I'm not going to stop you.
And you can get out the same way you got in you know,
in case you're wondering,
just roll out the other window!

Which, as you can see, is equally wide and equally open.
And well, ciao!
Have fun.

Song of the Flour Mill

rollers spin inside my body
driven by belts of breath
watch it fellows
mind your turbans

whoever you are, woman,
brahmin, ghatan, whatever,
makes no difference
tuck your sari and scram

flee, go home to your momma
I don't even want to know
whether you're a young woman
or an old bag

whether you're dagdu, thondu or pandu
a man, a woman or a eunuch
black, white, fat or thin
is all one to me my friend

but do yourself a favour
drop your jowar
and run
run for your life

back off boss
and keep your distance
unless you want to end up
in a sack of flour

it's gone kill-crazy now, this mill
and it's not going to be satisfied
until it has ripped
every hand that feeds it

chugchug chugchug chugchug
the mill swallowed the mother of us all
swallowed her whole
in just one gulp

and what did she taste like
sweet? bitter? sour?
the mill doesn't know
and the mill doesn't care

A Prostitute on a Pilgrimage to Pandharpur Visits the Photographer's Tent During the Annual Ashadhi Fair

Hey Mister, you the photographer?
I want my picture taken with Vithoba and Rakhmai.

Rakhmai to my left, Vithoba to my right
and me in the middle. That's the way I want it.

Move over Rakhmai, step aside.
Make room for me between the two of you.

Oh you're hopeless! Why are you so stiff?
They're all going to say you're a plywood god.

Come close to me, Vithoo my dear,
and put your arm around my shoulder. There, that's better!

Rakhmai, you're jealous! But you don't have to worry.
I'll return your Vithoo to you before I go back to Bombay.

You'll fill in some nice colours in that picture, won't you
Mr Photographer?
Paint my sari blue, and blue the sari of Vithala.

I'll take a quick look around the fair,
go for a spin in the giant wheel,

pick up a good blanket for myself,
take a peek down the Well of Death if I find the time,

and be back again in half an hour to collect my picture.
You'll keep it ready for me, won't you?

Old Newspapers

Beware of the old newspapers
stacked
on that little three-legged stool over there.

Don't disturb them.
I know it for a fact
that snakes have spawned in between those sheets.

Don't even look in that direction.
It's not because of the breeze
that their corners are fluttering.

It's alive, that nest of newspapers.
Newborn snakes, coiling and uncoiling,
are turning their heads to look at you.

That white corner has spread its hood.
A forked tongue
shoots out of its mouth.

Keep your eyes closed.
Get rid of the whole goddam pile if you want to
in the morning.

The Fuse

A rotten smell was coming from somewhere.
I gagged and reached for the handkerchief in my pocket
when my little finger dropped to the floor.
I picked it up with one hand
pressed the handkerchief to my nose with the other
when my nose came off. I wrapped it up
in the folds of the handkerchief and stuffed it in my pocket.

That rotten smell was still very much in the air.
The nose twitched in my pocket.
I thought I'll take a closer look and see
if there were any maggots in my little finger.
That's when the fuse went.

KUNWAR NARAYAN

Six Poems

Translated from the Hindi by Daniel Weissbort and the author

The Key to Success

They both carried loaded pistols,
were afraid of one another,
the hearts of both
filled with an old enmity.

At that time there was only these two.
But when the shots were fired,
a third was killed—
he was standing at a tea-stall . . .

Then a fourth was caught,
who was not even at the tea-stall,
but at home, and on his testimony
a fifth was grilled whom a sixth,
identified by a seventh, had implicated,
an eighth being tried
and a ninth convicted—
the tenth, who got off scot free,
fell to his knees before the eleventh.

The latter
had found the key to a new success,
which he nonchalantly twirled around his finger.

Horoscope

Lying in a field of flowers,
I have often thought about the dew,
fluorescent dots
dripped onto the petals
with nibs of light.
What astrologer designed this complex horoscope
of the glittering firmament?
And why do these luminous signs vanish,
from one to zero?
Whose is this cynical, geometrical yawn?

And then I thought
about the bedraggled leaves under the trees—
Who thought up this mathematical puzzle ?
The wind is counting:
it gathers leaves anywhere,
and deposits them anywhere.
At times, it snatches a few leaves from the tree,
crumples them and tosses them carelessly away.
At times it spreads out a new sheet,
and doodles its trees, trees, trees . . .

*Ibn Batutah**

In the forests of Mabar, Ibn Batutah wonders:
Who are these people, impaled on bamboo stakes,
blood dripping from their bodies?

* Ibn-Batutah (Mohammad Ibn' Abdullah Ibn Batutah, 1304–1368), the greatest of Muslim travellers, whose journeys extended from the Niger basin to China, was born in Tangier. He visited India in 1333–34. The eerie incident relates to one of the many adventures he had on the Malabar (old name Mabar) or Coromandel coast after a shipwreck while on his way to China.

Such a darkness by day,
or is the Sultan blind,
through whose sightless eyes
I see a tormented page of history,
fluttering in the sick light of torches?
Who are these men, women, and children,
already half-dead,
their hands and feet torn off?
Infidels, human beings!
And who are these others, seated nearby,
getting drunk, contrary to all the laws of Shariat?

No, there is nobody, nothing,
it's all a nightmare,
this is not happening, not now.
It happened a long time ago,
in a time of savagery, of bestiality.
No witness, I . . . King,
permit me to withdraw.
It is the hour for my prayers.

Description of the Missing One

Wheat-coloured, a peasant's ways,
scarred brow,
height not under five feet,
talks like he's never known grief.

Speaks with a stammer.
If you ask his age, he'll tell you—
several thousand years, give or take a few . . .
Seems crazy, but isn't.
Has fallen off high places more than once,
and got all broken up, so

looks glued together,
like the map of India.

To the Mound

A few ancient coins in the pocket,
reins in hand,
struggling over rough ground,
on our limping ponies—
the mound where we camped, it's said,
was the point at which Alexander
tried to turn home.

'Home!'
How often have I caught the fretful echoes of that word,
in places where
conquerors are supposed to have come
to the end of their tether.

But we had reached the point by a roundabout route,
and quite by chance. That is, we'd not intended
to be standing there, in the midst of a bunch of disinterested
    school-kids,
surveying this desolate place,
wondering about Alexander's past, or his future:
'We can't go on,
we're tired, we want to go home . . .'
All together, they were on the point of mutinying.
Had I been Alexander,
maybe I'd have done away with the lot,
all thirteen ringleaders.
We were within an ace of reaching that point
where Alexander turned back.

What a Day!

I was out all day
and nothing went wrong.
I met people all day
and wasn't insulted.
I told the truth all day,
and no one felt bad.
I believed everyone all day,
and no one took advantage.

And the biggest surprise of all—
He who returned home in the evening
was myself, intact.

MANGALESH DABRAL

Eight Poems

Translated from the Hindi by Arvind Krishna Mehrotra

The Death of Leaves

The leaves that settle on my face
Fall from my childhood's trees.
A lake sends me its waves, and
Like a wave, night shivers. I walk
On it, the death of leaves on my face.

The birds have spoken their lines.
The roads are empty; their lights
The colour of ash. The houses on either side
Have locked front doors. I call out
And my voice rebounds.

Good for a Lifetime

Perhaps there was a bit of moisture there
or a pastel shade
Perhaps a shiver, perhaps hope.

Perhaps there was just one teardrop there
or, as a keepsake, a kiss
Perhaps there was snow there
or a small hand
or the attempt to touch.

Perhaps there was darkness there
or an open field
or standing room
Perhaps there was a man there
in his own way struggling.

Woman in Love

The woman in love
Has this dream every night.
What's it about? One morning
She decides to find out.

She sees the commonest things around her:
A stretch of sand;
A tap left running;
Her disarrayed room;
And something she can see
She cannot see, though she looks again.

The woman in love
Trusts nobody.
She drops her comb
And turns her back to the mirror.
She says, I'm okay as I am.

One by one her friends desert her.
The sun goes down, keeping its distance.
The wind blows, but not through her hair.
The table is cleared
Without her having eaten.

The woman in love
Is deceived every day.
She doesn't know what's happening outside,
Who does the cheating, who takes her for a ride,
She doesn't know how it all started.

The world's a child in my arms,
Says the woman in love.
She comes out on the road alone
And looks at the big city around her.
Somewhere or other, she says, I'll find a place to live.

The Quiet House

The sun by slow degrees heats up the walls
Somewhere near, a slow fire burns
There's a ball lying on the bed
Books, the storehouses of disaster, are silent.

I'm half awake, half asleep
Half asleep, half awake
Listening to outside sounds
No sobbing in them
No threats being made or fear expressed
Nobody praying, nobody
Asking for alms.

And no bitterness in me
Just empty space waiting to be filled
And easily inhabited
Nor do I feel helpless
But an aching spreads through my limbs
And I recall the house of my childhood
Its back yard, I lying on my stomach
Taking the sun.

I ask nothing of the world
And can live as squirrels do
As grass does or a ball
That any little push
Can bring this quiet house down
Doesn't worry me.

Poem of Dreams

There's no running away from dreams; they happen as a consequence of waking. They tell us what we were and what we shall become. They make our half-lives whole. While we breathlessly rush about one hemisphere, they keep us quietly asleep in a corner of the other.

In dreams the earth looks round, just as our schoolbooks had said. The sun's heat is intense and stars shiver in their cold light. Trees of happiness grow around us. Someone on a bicycle goes by; we hear a radio playing. We see our roots immersed in clear water. We see the moon shining in a small dark room.

In dreams we see ourselves as righteous men. We see an old cracked mirror. We see blood coming out of our nose.

Poem of Paper

One day we find sheets of paper that were once important lying everywhere around us. We see them even as we go to sleep. They put an end to our dreams and cause insomnia. Much as we'd like to, we cannot sell them to the ragman, for in them our everyday lives are buried, those things we hesitate to admit to ourselves. Instead, we have to sit down and destroy them.

This is how old letters get torn, written by sympathetic friends when we were down and out. Declarations of unrequited love, along with the poems of some major poets, words we believed would remove the world's hunger, get reduced to shreds. This paper now won't make a child's boat or his missile even, the kind that goes a short distance and turns back.

We've become wordless, and all but lost our speech. We go on tearing the paper. It's our only hope.

Words

Some words scream.
Some take off their clothes
And barge into history.
Some fall silent.

*Outside**

I closed the door
and sat down to write a poem
There was a breeze blowing
and a little light
A bicycle stood in the rain
A child was returning home

I wrote a poem
which had no breeze no light
no bicycle no child
and no
door

* Translated by the author.

MIRON BIALOSZEWSKI

Ten Poems

Translated from the Polish by Daniel Weissbort and Grzegorz Musial

Secret Freedom

They passed on to one another
walls like the plasm of life.

In the family they raised
        great-grand robes
        great-grand knobs.

When the stairscrolls unwound,
when the house sank
                                        to the bottom,
they felt: the wardrobe's still a pain in the gut,
they almost thought: naming it ensures
                                                        the sideboard's existence,
        just let out a croak—
        and the door's laughing . . .

Through a landscape broken up into syntheses
they ascended to the promised floor,
which each
had promised
himself.

They didn't chant Ecclesiastes.

Vanity is for the chosen.

The chosen are few.
Because—each one
        chooses—
                an own self.

Dream Evidence

Behind fences of bedsteads
we—cinema shacks
of dream.

We cannot stamp
nor can we clap.

The most is
we can scream apelike
in our ancient dialect
of the latest items.

And really
that's when we experience
our own civilization.

How Easy to Lose Faith

A horse and haycart passed.
I see. I believe in them.

It is growing dark.

A horse and haycart passed.
But the horse had a horse.
The haycart had a haycart.
They led those things of theirs
great shadow things
through the acacias.

And it's not so easy now
to believe in horse and haycart.

Self-Portrait in Joy

Don't think I'm unhappy.
I rejoice that I think.
And you, think I'm rejoicing!

Consciousness is the dance of love.
My consciousness dances
        before the lamp of rain
        before the shell of the wall
        before the grocery store with its assemblage of cabbages
        before the mouths of talking friends
        before my own hand unexpected
        before the unhollowed out moulding of reality—
in opulence at the height of the celebrations
and the loftiest service
inseparably
my consciousness dances.

And when the dance unravels,
as skeins will,
I shall go to heaven—
            where one feels nothing,
            where from the start I was before I was,
            where I shall be until the end, when I no longer am,
            there—more joy than you can say.

. . . .
That's all.

Self-Portait as Felt

They look at me,
therefore it seems I have a face.

Of all unfamiliar faces
I remember mine least.

Sometimes the hands
live for me utterly apart.
Shouldn't I then add them to one another?
— — —
Where are my boundaries?
— — —
Yet I am grown over
with motion or half-life.

Always, however,
there creeps in me
something,
full or unfull—but existence.

My self
brings along
an own place.

When I lose it,
it means I am no more.
— — —
I am no more,
therefore I don't doubt.

In the Window Pane

The square undershines me.
The street has a vista, it comes
goes, one shines, a second, a third
turns
trolleybuses turn away from the vista
obliquely squarely
someone is walking holding a head
on backwards in white
a babe in white
on one side they're building
                                        details
                                        details

to sum up—I—stand
—in the window—in the dark
not yet six—and what background?
—I wait—
—like painted—

Monklike Cynic Listening
or
Thin is the Rainbarrel's Skin

in the distance
                    the rain
                    objective
                    stands
                    pyramidlike
and right here
                    the ears
                    drunk
                    on uniform gutterpipe music

the truth

drips on into the barrel

it's where
dwell
the Diogeneses
of my ears
minimalistic

My Fear Non- and Organic

Last night I broke my comb.
At night I was scared
        of the guardian-rooster
        and the rusty saw,
was afraid of my squeaking stomach,
which I took for an invisible mouse.

*Study of a Key**

A key
smells
of nail water
tastes of electricity
and like a fruit
    it is tart
    unripe
    entirely self-enclosed
    a seed.

Quatrain of Self-Speech

one hair is beautiful
I have a patch up there
myself still unpatched missing
beauty by a mere hair's breadth

* This translation first appeared in *Poetry East* (Chicago).

GRZEGORZ MUSIAL

Eight Poems

Translated from the Polish by Richard Chetwynd
and the author

I Know How it Begins

I know how it begins
I saw the fear on my father's face
unconscious yet
there was understanding
the mouth the eyelids the fingertip
on the topsheet

I watched
I wanted to remember
the book the glass the thermometer
the rustling birch behind windows
I know how it begins
the leaf breaks from the branch
and sits awhile on the windowsill

VIII 84

Only a Shadow

between the tile furnace and the balcony door
at the crossing of water veins
I turn from side to side the shadow pulls my hair keeps me
from sleep;

it's her
slippers clicking coffee shaking in her hand
God promised eternity
now he kills
tears down the wallpaper spins cobwebs
a maniac
her graying head
he struck against the edge of the couch
a drop of saliva dripped from her mouth;

who will restrain him
he takes back everything he gave
the taut muscles sweaty skin the sunburned neck
castles churches the car passing over green hills
he murdered father and
now her
even Schubert's yes 'The Linden Tree'
she played
leaning by the piano
sister turned the sheet music;

between the tile furnace and the balcony door
I lie in a dark room listen to underground streams
Elizabeth Schumann on the worn-out record
even into the singing larynx
he pours ash and sand

nothing more
the faded photo of Rimbaud

a shroud on the wall
then Meleager rubbing olive oil into his hands—
go back
among the shadows
the trees stand in frost
a branch scratches the glass
the mute record smells the light of the grave

XII 88

The Accidental Witnesses of Events

the accidental witnesses of events
by accident happen to be here
they are witnesses of events
beyond their measure

the accidental witnesses of events
will make of them too much or too little
seldom striking the core of hope or bitterness
but only they'll remember the color of the victim's shoes
whether the ring was still on his finger

don't ignore the accidental witnesses of events
only they will say
who supports the trumped-up king
how many hands reach for his crown
and who behind the stage takes careful notice

Gdansk, VIII 80

Letter to Paris

we French won't be dying for Gdansk

we people of Gdansk
will be dying
for the Louvre

for the finest linens
and china
for Caruso records
and Madame de Stael gowns
for Arras carpets
and boys removing thorns

we people of Gdansk
remember
the walls cracking
in flames
and the silverware
cast into hell's pit
inside the gilded coffers

it's happened once already
that we gave
for the usual reasons
why shouldn't the usual
reasons
again
claim us

Gdansk, X 81

1939

Above all saved will be those
who are guiltless idiots
little hussies in silver blouses
who understand nothing

Who ran from the dance beneath
bombs hugging useless
handbags the most realistic

trickle of tears on soot-smeared cheeks in truth
it is not they who invented darkness

The little furry beasts on high
heels the graves
without names and dates

Suicide in the Hotel

her leg slipped to the carpet
her face clings to the pillow
you could lift her stiffly by her hair
from head to toe—

you could appeal to her conscience
you could pound with your fists
try moving her with a photo of her children
you could mock or cry from heroism to idiotism
run through all the words it's too late
to gain on her

she smiles evenly
a drop of saliva
the eyelids—
the sentence 'i
can't' that's it
there's nothing
more

That Place

that place.
here I grow. here I sing.
here I try. here I lose.
that place. not the table. not

the chair. not even
the house.

these people.
here they trust. here they wait.
at these windows they look for me
everyday.
familiar coat. familiar gait.
on this door everyday I knock.

that place
which denies me
the immensity of other places.

The Answer (II)

this brightness in me
do I finally understand? no
since yesterday it's been difficult
to define what's sealed within
me the fatal
imprint I saw in the sand
while walking the highway I shouted
O

(and hidden
in the shadows She
stood behind trees
and laughed
silently slow-
ly)

IX 84

LIA PURPURA

Some Notes on Translating the Poems of Katarzyna Borun-Jagodzinska

Had I known what, precisely, I was getting into when I decided to translate as many of Kasia's poems as I could between September and December 1989, I am certain the task would have appeared overwhelming and the problems, thankfully unforseen, altogether baffling. I'm still amazed at the roadblocks I've found en route from Polish to English, but more than that, I'm interested in them and amused by the variety of ways they present themselves. Perhaps this is the most fruitful way to enter into any new territory—blindly, somewhat dumbly and blessed with a kind of momentum, bravery, or cockiness.

Translating from Polish and not knowing Polish is, of course, a difficult position to defend, but the strengths we had in our favour seemed to outweigh the barriers: Kasia's English is pretty good and we're both poets. Her syntax is straightforward and her poetic subjects and stance toward the world are clearly argued. In other words, she takes a stand and each poem has a mission, a journey to complete. Of course it would be ideal if I spoke Polish but since translation seems to require such a leap of faith anyway, I suppose our method included a good deal more faith and a lot of leaping in the place of certain assumed basics.

Our method worked like this: Kasia first read me the poem in Polish and I listened for inflections, tones, rhythms, rhymes, any aural guideposts. Then, starting with the title, we moved through the poem line by line. This line by line advance seemed to be the most natural approach for me and one that most closely approximated the act of writing and discovery of the subject while writing. It never occurred to me to want to know what the poem was about before starting the translation. Because Kasia has a perfect ear for idiom, which phrases might require longer discussion, we stopped often to wrangle out what I began calling the ten poems we needed to write together in order to get to the one on the page. Learning the tangentials was perhaps the most crucial part of translating in the right spirit. I imagine it was difficult for her to talk in such detail about the work's origin and each image, and risk flattening the poetry of it. After all, no artist takes kindly to explaining her work down to the bone. When we reached the end of the poem, I'd read the first draft over to her, ask more questions, take more notes and we'd go on to the next one. Then we'd meet in a few days and go over everything again.

While it's certainly difficult to 'experiment' with one's translations under the nose of the living writer, I found it was possible to stray further from the absolutely literal text as time went on and as a certain trust developed between us. The writer's trust and flexibility is tremendously important. For instance, the Polish version of 'he missed the boat' is 'he missed the train'. The idea of motion, of missing the mode of escape was important to this particular poem, but I couldn't make 'missed the train' carry the weight of the idiom she had intended; it sounded too much like a specific incident and not a concept—luckily, after hearing my explanation, she was willing to barter one *train* for one *boat*. Though her work was so clear and the logic of the narrative was so lucid and didn't take too many odd twists or stream-of-consciousness liberties, this very clarity often made it difficult to 'turn around'. On many occasions it seemed impossible to stray far enough to make poetry out of the literal. I often felt I was driving a finely tuned machine that was so responsive, any slight nudge or tap on the steering wheel would throw the thing wildly off course. So we had our share or arguing, convincing, cajoling.

Because her subjects are particularly 'Polish' (living in the time of martial law, etc.) I had to learn to think of 'history' from the inside, which means it becomes smaller, more a daily occurrence. History for many Americans my age is newsreels, the newspaper and in my case, thankfully, immigrant grandparents who are still telling their stories. History, for the Polish, is a tank rolling past the window. How can I translate 'prison', I thought, let that word loose in America and hope it means what it means to her. Or even 'printing press', 'unfulfillment', 'longing', 'betrayal'? In Polish, those are public words that define a public and cultural psyche.

KATARZYNA BORUN-JAGODZINSKA

Eleven Poems

Translated from the Polish by Lia Purpura

The Match

There were so many of you
and you each cast four shadows:
one recalled the taste of your first smoke
the next your first swig of vodka
the next your first love
the next your first death.
There were so many of you
each at the center of a cross:
the first remembers the dizziness after a smoke,
the second the burning throat,
the third, a woman's sweat,
the fourth, the sickening smell of a corpse.
There were so many of you
and each shadow quadrupled,
a quadrupled soul that in the floodlights
lay itself down in the grass.

'Women Will Not Judge Me'

– from the diary of Kazimierz Raton

Only on the marble of the cornets
of the sisters' of mercy
will damnation break.
Only there will the blasphemies
of the dying flow down.
Women come when it's all over.
They will not judge.
Let us believe
they alone
wash and dress the dead.

The Night

I heard it clearly—behind the wall
they threw something into the garbage chute,
a heavy, painful fall.

It might be already-dead books,
still-living pups,
a baby wrapped in a newspaper,
even a dismembered corpse . . .
it might be, but in our house there is no garbage chute.

Her Steps are Softer and Softer

Softer
than your child's breath.
So don't try anymore
to name
what sneaked along the wall
and vanished around the corner
like the agent
of a foreign power.

Conversion

To Father Jan Twardowski

To return
means, of course,
you've been here before.
Home is near
but someone missed the boat
or took the long way around.
To return
also means
to gaze at angels
obvious as hummingbirds
for all of twenty years
and only today notice
the loose strap
of a tiny sandal.

Statement on a Discussion of the Apocalypse

We don't tremble for the fate of the world
like our older friends.
We know,
we know perfectly well
it will be folded
like a tent
pitched at roadside
for one night.
Let's only agree
we'll hold hands
tightly.
It's going to be wonderful—
all of us thirty-odd years old
and forever the youngest.

Photograph of Teresa Martin as the Maid of Orleans

And here
I am the little Joan.
Well—maybe a little older.
A *tableaux vivant*—
armour, lance
and a smile tough as iron.
Yes
a real-life *tableaux vivant*.
Perhaps I don't yet sense
I am slowly climbing
the burned stake of my body.
(Now—we'll have no more talk about the body
unless you're referring to martyrdom)
Perhaps I already know
about the pile of ashes
and the indestructible lump of gold
they will find there.

Teresa in the Courtyard, a Month before Her Death

Everything is in Your hands.
But I'm not a fly.
And You're not a naughty boy.
I know, after all, this darkness
that creeps toward me
is only the shadow of Your smile.

Last Photograph of Teresa Martin

Before someone comes
with magnesium on a stick,
before they set a camera on a tripod
you'll all be walking softly,
talking in whispers,

afraid I'll run away,
bound off to endless meadows.
I can see I am fading
with your help
like a gaudy print.
You will be walking softly
and whispering
afraid of startling
the angel of death.
Though he is
the least timid of us all.

Iowa City, October 1989

It will be banal, I'm sure. New York
'empty hotel room
four in the morning . . .'

No, that's not it.
Dormitory,
two a.m.
Iowa City.
Joe Cocker and his groupies
in the corner, squeezed like a Panasonic radio
which has to drown out
someone dreaming behind the wall.
The young man who shakes
like someone with Parkinson's,
is only on film.
The heavier, older version
sings, overwhelms us as well as he always did.
Though Woodstock and whatever else
is twenty years behind him,
perhaps it's gallons of whiskey.
He sings.

America.
It's like
after many years
the cold kiss
of lovers who have never loved.

Butterfly

– for Krzysztof Piechowicz

Ah, Madame Lilka*
don't cry—
you are nothing like a butterfly,
not even a chrysalis,
naked as an earthworm
against the backdrop of a rainy night.
Ah, Madame Lilka,
that's no heart beating,
only gongs.
The orchestra is out of tune.
Her lord picked a mulberry leaf,
crushed it between his fingers.
Green stains the white uniform.
The wind knocks down the screens
of rhyme.
Take the child away
don't let him see this.
From the tree, sticky black kisses
drop down.

* Lilka: the nickname of the great Polish poet Maria Pawlikowska-Jasnorzewska (1891–1945)

JAMES S. HOLMES

Studying Translations: An Underdeveloped Country in the World of Literary Studies

James S. Holmes was born in Iowa in 1924. He settled in the Netherlands in the late 1940s, becoming the leading translator of Dutch poetry, for which he was awarded the Martinus Nijhoff Prize, in 1956. In 1984 he received the Flemish Community's first Triennial Prize for the translation of Dutch literature. In 1960 he became a lecturer at the Department of General Literary Studies of the University of Amsterdam, where he pioneered translation studies. He was in fact the doyen of European translation studies, and his importance in this field is brought out by his recent, posthumously published book, *Translated! Papers on Literary Translation and Translation Studies* (Amsterdam, Rodopi, 1988). Besides being one of the foremost poetry translators and translation theorists of our time, Holmes was also a poet of unique candour. The talk, printed here for the first time, was delivered to a small invited audience on the occasion of one of Holmes' annual visits to his home State. The translations of poems by Sybren Polet are among the very few finished translations by Holmes yet unpublished. Jim Holmes died in Amsterdam, in 1986. (*Daniel Weissbort*)

I'm not one for formal lectures, and anyway for a group this size I think it would be rather ridiculous for me to give a formal lecture, but I'd like to talk to you, and talk with you, about a problem which I think is a serious one in literary studies in general. What I think I'm going to do tonight, since we're such a small group, is to try to state very succinctly what my problem is, and I won't talk too long, and then we can discuss it together and see whether we can't as a group get further. There are several comparatists here, and a number of people the comparatists sometimes disdain, called philologists and so forth, and several translators, I presume, and people with different interests in literature, so perhaps we might get further with this problem than I am at this moment.

'Studying translations, an underdeveloped country in the world of literary studies or comparative literature or something of the sort' . . . I am constantly struck by the fact when I look at any survey of a literature, a literary history—well, if we restrict ourselves to the literatures of Western Europe in the period at the end of the Middle Ages—that there is a great deal of attention paid to translations, and then suddenly they drop out of the picture. There is from then on talk about influences, the way that Shakespeare exerts an influence on German grammar, or Goethe on early nineteenth-century English literature to Coleridge, and so forth, but the word translation almost never appears in these further discussions except for the period when translation served a very important function as a way of developing the vernacular languages, and that applies to French, English, Dutch, and German very clearly in the sixteenth–seventeenth century. This constantly amazes me because how do literary influences get across linguistic boundaries? There are only two ways. One is that the writer in the second language reads the first language, and so gets to know an author in the original and so is influenced by him, and the other is that he reads a translation and is influenced by the translation. In some cases it is clear that it was the first situation. The influence of classical literature on modern writers up to the Napoleonic days was probably primarily through reading the original language. But I would hesitate to say that that has been true since then, as far as classical literature and Western writers are concerned. And it has never been the case, for instance, with Russian literature, which has had a tremendous influence on modern literature in the West. Of course, there are exceptions like Nabokov, but most of the influences of Russian literature have come into the West via translations. Yet you would almost think that the literary historians, the literary critics, who attempt to develop a larger than national view of literature and the way literary currents flow, have assumed that a translation is a perfect mirror of the original, such that one can ignore that link and simply say that Tolstoy or Dostoevsky became an influence on the contemporary novel, ignoring the fact that it is Constance Garnett's Dostoevsky that is the influence. Now this is surprising because when literary people do write about translations, contemporary translations, they are very quick to point out where the translations differ from the originals. I think there is a major problem here, which is that in the literary world we are largely still very naive in our thinking about translations.

You can see that, too, in the critical writing about translations. I don't think there is any area of literary studies where there is so much naiveté exposed by serious scholars as when they write about translations, when they review translations. They immediately begin with such remarks as 'translation is impossible,' which is begging the question. I think that it is simply a question of how you define the word translation. If you say a translation should be a perfect mirror of the original, then translation is impossible. But if you say that translation is at the level of adequacy, presenting a text which has a great many connections with another text, then translation is obviously possible. But at the same time translation is transformation. It's producing something different from the original. Critics seem to be very much concerned with a problem which I suspect is not actually a problem that can be solved in a scholarly way, that is, the quality of translations. They're constantly evaluating whether this translation is a good one, whether this translation is bad, and so forth. Most of the reviews, in fact, praising or condemning translations without giving grounds for their decision, are simply impressionistic evaluations.

Well, it would seem to me that through the years the fact that translation is an imperfect instrument, that translations are not mirrors of the original, has led to a great many distortions in the recipient culture. To cite an instance, which is the instance I used yesterday in the workshop, Western literature has since the Renaissance looked to Catullus as one of the leading love poets of the classical world and the father of all passionate love poetry. It has been able to do that only by ignoring a large part of the oeuvre of Catullus, so that those few passionate love poems to Lesbia have been given a prominent place. In the biographies of Catullus, about whom we know practically nothing, pages are devoted to this Lesbia, mostly distilled from the poems, ignoring the homosexual poems, ignoring the obscenities, ignoring the vitriolic satire of Catullus. We have been presented since the Renaissance with a picture of Catullus which is not the true Catullus. Now it is not the Catullus scholars who have done this, because the Catullus scholars until quite recently wrote in Latin, like Catullus, and wrote for other Catullus scholars, so that they could talk about all those things that you did not talk about in the vernacular in the Christian, Western world. But the Catullian tradition in the West, the way he is presented by the translators to readers of translations, distorts the picture. This may be an extreme instance, but there are very many similar, if more subtle, cases such that I think for my generation—and again I'm going back to the example of the Russian novel—the sense of nineteenth-century Russianness in the novel to a large extent reflects the emotional life and style of Constance Garnett. And it seems to me if we are really going to try to analyze what is happening in the development of Western European literature from the Renaissance on, we will have to place translations in a much more central position, to see how influences get filtered, shifted, obscured, twisted, skewed by translators.

Now, one of the problems is that we have no methodology for studying, for looking at, translations and describing how it is that a translation differs from its original, and what happens in this translation. [. . .] Maybe the next step is to start talking with each other about what happens in translations, and what has happened in translations in the past, so that we can see how influences have been filtered. This is something quite different from the evaluative, normative analysis of translations that I was talking about a few moments ago—whether a translation is bad or good—since every translation is a bad translation if your standard is that it has to be exactly the same as the original. Every translation is the product of a translator trying to look at a text and to solve problems about what can be done with this text in another language, in another culture, and each translator is going to solve those problems in a different way. It depends on his own personality. It also depends on the recipient culture, the language, the time, as to what type of problems can be solved. I have no quarrel with Catullian translators of the late eighteenth century who made Juventius (more the Catullus, or at any rate the poetic 'I', who was in love with making Juventius) into Juventia. It's quite understandable that it happened. But it is something that analysis should bring to the fore, and it says something about the recipient literary culture of the time. Likewise I have no quarrel with the fact that Latin poetry was translated at certain periods and in certain countries in a sort of imitation Latin meter and in other countries and in other periods, into native forms—and in our day into a sort of free verse. But we do need to be aware that these things are happening and that these are translator's solutions to problems which need to be brought to the fore.

I think that, quite aside from developing a methodology, it would be very wise if the translator more than hitherto—and if publishers more than hitherto required it— described what he has tried to do with the text. For instance, if he's translating Catullus again, in some form of free verse, he should at least describe the original meter, so that the reader would know that meter is there and that it doesn't function in English, which is why the translator has done this or that. Or with certain imagery in the original—there is that classic case in Middle Asian literature where it is highly lyrical to describe a man's face as gleaming like a leech, and where the image is obviously not translatable in this way and with the same effect into Western culture—the translator should alert the reader, so that the reader can see what is happening to the text when it crosses linguistic and cultural boundaries.

In other words I'm pleading for more sophistication in the use of translations and more sophistication in the study of influences. We need to know much more about what is North's Plutarch as opposed to Plutarch himself, or Florio's Montaigne as opposed to Montaigne himself, and Dryden's *Aeneid* as opposed to the *Aeneid* itself, and to what extent the influence of Virgil on English in the eighteenth century was Dryden's and not Virgil's. We can only do this when we place translations much nearer to the core of literary studies and not out here somewhere on the periphery that some offbeat character writes a thesis about.

That is what I wanted to say, and I've put it in a nutshell instead of going on at length. I hope that I've been radical enough, extreme enough, for some people to challenge me.

QUESTIONS

[Many of the questions are inaudible, though they could perhaps by reconstructed from Holmes' responses.]

Q: [. . .]

A: A student of mine told a story a couple of years ago—a Dutch student—he'd been reading *The Brothers Karamozov*. This was more than a couple of years ago because it was in the day when [???] Park in Amsterdam was the center of European freedom. He was wandering around the park and he met a man who was also reading *The Brothers Karamozov*, but he was reading it in English. They got into a deep discussion and they discovered that there were a great many different impressions about the book, and almost all of these were reducible to the differences between the Dutch and the English translations they were reading. It is not so much a question of the translation of a particular word or a small detail, but the general approach to the book, the general style of the book, the general sweep of the book, and neither of the two had really thought about translation before. They felt, this is a Russian book I'm reading, I'm reading it in Dutch or I'm reading it in English, but it's the same book. After the discussion they came to the conclusion that translation is such a rough filter that it was really two different books they were reading.

Now that's going pretty far, but it seems to me that there is a terrific amount of naiveté in this area. You ask a student who's reading any translated book who translated it, and chances are he will never have noticed the name of the translator, that this is an element that is so unimportant, that he feels he is reading the original—in French or in English, but the original. But I think it is very important that we make people aware that it is never the original, and that in maybe five years from now another translation will come out which they should immediately read because it will be much better than this one and much closer to the original. For instance, I remember when I was young I read Greek tragedy in the translations that were available at that time, like Gilbert Murray, and what a revelation it was when those new translations appeared during the war, contemporary translations. It was a completely different world than the one I was introduced to before. At any rate, I think students need somehow to be more aware of the nature of translation.

Q: Do you think that each time a book is translated, or let's say retranslated—because each century makes different demands, so the same books need to be translated over and over—the translator might very well in the introduction point out why there is a need for a new translation? Is this done very often? I don't see it very often.

A: It's done a little bit with poetry translations, it's rarely done with prose translations. And I'm afraid that the publishers are the cause. They don't really see the need for it, and they certainly wouldn't in general want it in an introduction to a prose text. They might with a great deal of pressure permit an afterword. It is understandable, because the publisher wants to maintain the illusion that this is the original.

Q: [. . .]

A: A translation does not convey everything that is in the original, but I always object when someone says that translation is always less than the original. Translation is always different than the original, but sometimes there are brilliant things in the translation that are not in the original. It's something else, it's something different, but it can be just as good as the original. It seldom is, but this is a possibility[. . .] Most of us feel the translation should not be better than the original, but that was often not so in the eighteenth century. Many people in the eighteenth century thought they could improve on ancient Greek texts. But obviously the translation is going to be something different.

I think there is one exception to this need for new translations. Sometimes translations which are almost contemporary to the original can last. For instance, I don't think any of us would want to read a nineteenth-century translation of Rabelais, but I think the seventeenth-century translation in English is readable in the same way as the original. [. . .] Translations either have to be contemporary with the original or contemporary with us, and nothing in between satisfies, although this is not always the case. There is a Dutch translation of Whitman contemporary with Whitman, and the Dutch is absolutely unreadable. The Dutch language has changed so much over the past hundred years that in this case the translation is absolutely unuseable. But I think I missed part of your question.

Q: [. . .]

A: Translation always says something different from the original, and therefore it cannot take the place of the original. However, there are a few cases where a translation *has* to take the place of the original, when the original is lost and only the translation is left. This is a theme that crops up now in science fiction, that all of the culture of our day and our past is preserved for a far distant future only in the English of this time. It is not so inconceivable. Then you would have to make do with translations which would take on the functions of the originals. But in general translations are there because people cannot read the original, though I think that there are two different kinds of rough filter. One is the filter of the translation. The other is that, say, of an American who's had three years of college French, reading a novel in the original French. That is a very rough filter, too, and it is problematical as to which of the two brings you closer to the original: reading it through the filter of your imperfect knowledge of the language or reading it through the filter of the imperfectness of the translation? It's a choice of two evils, but that is the way that literary influences with all of the distortions that take place have spread across languages, in general. And I don't think that it is a bad thing, actually. You see it, for instance, in the influence of American literature on French literature, the rough filter, which is partly translation and partly an imperfect understanding of American culture in France, which Frenchifies all American influences to the extent that they are almost unrecognizable [. . .] French Chandler-type detective novels, somehow are very French and not very tough-guy American.

Q: [...]

A: ... Popovich in Czechoslovakia has been working in this direction. It's nothing that any of my students have been working on recently. I am conducting a seminar at the moment in which we're simply trying to collect as many translations of Catullus as we can find in Italian, Spanish, French, German, Dutch, and English, from the Renaissance on. We have put together a small corpus of ten poems which are mostly homosexual, and some that are very obscene, according to traditional standards of obscenity in the Christian and Western world, simply to see how the translators solved this unsolvable problem—I mean homosexuality was something taboo up to the French Revolution and the obscenity was something that was taboo, to the extent that Catallus uses it, up to ten years ago—to see what sort of solutions translators would apply. I hope to write a little article about this next year [unfortunately he never did, but see *Translating Poetry: The Double Labyrinth,* ed. D. Weissbort, Macmillan, 1989], I think it's going to be a small book called *Translating the Taboo,* which will begin with a brief methodological article on our work. Then there's something else I want to examine. A Dutch poet, also dealing with homosexuality, who uses the second person in his poetry, which you can do in the Germanic languages, and it doesn't make the sex clear. But as soon as you translate those poems into the Slavic or Romance languages, you can't avoid making a determination of sex. How did translators solve this problem? . . . And there are some other examples.

Of course, every society, every recipient society, has its taboos. We have the feeling today that we are free of taboos compared with all previous societies, but probably it's because we're blind about the taboos we have. These taboos also influence translators. And even if they don't, they influence their publishers, who either don't dare to publish the book, or often—at least they used to—censor it. One of the great translation theorists, of sixteenth-century France, Etienne Dolet was burned at the stake because he translated Plato in an un-Christian way. Here was a translator who thought he was being faithful to Plato, who was after all not a Christian, and the Church felt that was such a sin that he was turned over to the secular arm in 1548, which was very late in the day. So, every translator is faced with these dilemmas, and that's what I'm interested in getting at in this Catullus seminar.

A student of mine a few years ago wrote a paper, a long paper, about six translations of a play of Plautus, and he simply went through and noted down every cultural item he could find in the play which was inherent to Rome or Greek culture in the classical period and which did not correspond to a cultural item in our own time. At the end he had listed two hundred and fifty items. Then he went through five translations, from the thirties to the fifties, to see what the translators had done with these. There was a great deal of variety. One translator retained, oh, eighty percent of them, and another retained something like twenty percent of them, but no one retained them all, and no one got rid of them all.

Q: [. . . .]

A: In the second-half of the seventeenth century and early part of the eighteenth century in France, translations—to use our traditional terminology—were very free, partly because one of the demands of French writing at the time was that all translations had to conform, as all original writing had to conform, to the very strict stylistical demands of classical French though at the same time there was a terrific controversy that Madame [???] was involved in as to whether you could introduce or leave in elements which were in the original but which were . . . well, it was just not done, at any rate, in better circles in France. There are many elements in Homer, for instance, in both the *Iliad* and the *Odyssey*, which were considered bad taste, and therefore were simply written out in many translations. We feel, when we look at Pope, that he did that, but Pope was felt in his own day to be much more faithful to the Greek texts than any of the French translators. Pope is a good example of someone who felt he was trying to get to the essence of the Greek text, whereas for us his translations now seem to be the very elegant work of an eighteenth-century Englishman in Greek clothing. It's quite possible that our translations will be looked upon, in the future, as distorted, reflecting our own day, our own culture.

Q: How do translations which have become part of the national literature, as Pope's have, relate to future translations, or translations of our time, which seem, to us at least, to be much closer to the original. I mean translations do typically have a limited life, yet some translations enter into the literature, and so the whole thing becomes extremely complex.

A: Well, I think many more translations enter into the literature of a particular time than we see when we look back, and that is part of my earlier argument—that we generally underestimate the role of translations in the whole life of a literary community at any given time. Now, in a very large literary area like the United States with a prolific production of its own literature it is somewhat less strong than, say, in Holland, where a large part of everybody's reading consists of translations of French, German, and English literature, and the literary work of the moment may very well be an American novel, or a German novel, or a French play or something. They're usually read in translation now which is quite different than twenty, twenty-five years ago, when a much greater effort was made to read everything—at least in the French, German, and English languages—in the original. But much of the literary life of any literary community consists of translations. Think, for instance, of American literary life of the last fifteen years, of the South American, the Latin American novel, which we all read—according to all reports—in excellent translations. But these tend to get lost in the literary histories. They will be picked up whenever there is a strong influence of one of these—Marquez, for instance—on some native American writer, but otherwise this vast amount of literature, which is so very important in the literary life of a community, is largely ignored in literary histories. [. . .] If the literary histories were rewritten in this way, then there would be many more texts, besides Pope's *Iliad* and Virgil's *Aeneid* and Chapman's *Homer*, and so forth, which would be seen to have played a major role. [. . .] Burton's *1001 Nights*, for instance, doesn't usually get into the literary histories, but I think that it is probably one of the great formative books of the modern era.

Q: I'd like to ask you a technical question. Let's say that you have a short text, which you analyze and discover to be extremely complex. There are so many facets that if you try to render, to be faithful to, all of them in your translation, something is going to have to give. What criteria can you apply in trying to preserve as many of those various facets as possible? Or is this a false problem?

A: Well, it's a problem which a number of linguistically oriented translation theorists have tried to solve, but I don't think any one has succeeded in establishing any sort of valid hierarchy. When you compare the theoretical hierarchies that are proposed with what actually happens in existing good translations then you falsify all their contentions. [. . .] A great deal depends on the personal abilities of the translator. I'm a bad rhymer, so if something has to give, it's usually going to be rhyme, whereas I have some friends who are brilliant rhymers and they would be reluctant to drop the rhyme in translating poetry. No, I don't think, it is a false problem. [. . .]

Q: [. . .]

A: There is one basic rule, I would say, that is valid for literary translation of a literary text, or a translation which aspires to be a literary text in itself. There are also possible, quite permissible, translations of literary texts which do not pretend to be literary texts. But if you are trying to translate a literary text so that it will function as a literary text, then I believe the interpretation of the whole must take precedence over individual parts. To achieve what your interpretation of that whole makes you want to achieve, you have the freedom to deviate in the parts, if necessary, and I believe—though I'm very skeptical of norms in translation—that it would be a violation of the whole to retain a part which didn't function in terms of the entire translation. This is one of the places where the linguistic translation theorists go wrong, because they see translation, basically, as the translation of a series of sentences, rather than of whole texts.

Q: Privileging the whole of a text, is there a point at which specific literary methodologies help? [. . .] Are these stages along the way in which different methodologies can help?

A: I have never been able to come to a classification or hierarchization. To return to the leech image I mentioned earlier, just imagine, in whatever Central Asian language this is, that there is a novel in which this is a key image, where it also plays a very important role. In that case, you cannot change the image, because it has this double function, and the only way you can solve this is somehow to take away some of the aversion to leeches by explaining its function, either in the text or in the introduction or in a footnote. Thus, one of the basic problems is that culturally-bound elements in an original text are usually homogeneous to the text. [. . .] Most American novels written in English are set in America, where you expect people to speak English and where the American reader feels—this is a novel about my environment, or, of course it's a bit exotic. Sitting out here in Iowa City, reading a novel about New York City, there is a certain exotic element, but still the language and the situation are homogeneous, function as native for the reader. There are also many heterogeneous texts, for instance, an American novel set in France written in English. There the language is no longer synonymous with the situation, and there are certain problems.

The strange thing about translation is that almost always a homogeneous text becomes heterogeneous, and at the same time it becomes exotic, novel. It's a text about somewhere else, and reading exotic and heterogeneous texts demands much greater sophistication. [. . .] Since the late eighteenth century we have loved to read about strange places and strange peoples, and one of the reasons that we translate novels is because they are not about ordinary, everyday life. They are escapist, in a certain sense, they take us to another place. Science fiction is another kind of heterogeneous text—all those people living five thousand years from now, speaking twentieth-century English. We accept this illusion, but there are limits to our acceptance. It depends very much on the art of the translator whether we're willing to accept an image like 'as beautiful as a leech.' I think that we're accepting much more now than a hundred years ago, and certainly much more than two-hundred years ago, when such exotic images would have been immediately replaced by acceptable English ones. But now we are intrigued by images which are not in our reservoir of images; they're different and therefore interesting and exotic. So, I think that translators can do that sort of thing much more now. [. . .] I believe that a study would probably demonstrate that translations tend first to be more adaptive, to bring the book closer to the reader, to make it more native. But once a book becomes accepted as an important book, a classic, then new translations will retain more of the original and be less adapted to the recipient culture. Because it has a reputation, you'll accept it whether you like it or not. [. . .]

Q: [. . .]

A: Schleiermacher advocates that the reader should be brought completely to the source culture, even to the extent that the German should reflect Greek syntax and so forth. I don't think that we accept that.

Q: I have a practical question. What you sort of suggest is that in an ideal world the teaching of literature in translations would be accompanied by a description of what happens in the process of translation, so that people would have a much better idea of what they're getting, how it relates to the original. Can you see any way in which this can, practicably, be done, say, in a university? Or do you think that a lot more material, a lot more critical work on the key texts is necessary?

A: I think that a great deal would be accomplished simply by little notes about the translator. [. . . .] What I would prefer, of course, is for the translator to write a paragraph of his own.

But at any rate enough should be done to make people aware that there is a translator there, that he has a name, that he died thirty years ago and so he's not writing the English that we expect, and that his English, of 1920, is not the Greek of Aeschylus, and that sort of thing. It seems to me, well, in those early courses in the freshman and sophomore years of university just a few remarks here and there by a sophisticated teacher could do a lot, because at that level . . . well, I encounter again and again students who are simply not aware that there is a translator, but imagine that they have Greek texts which somehow are in English, and that this is the same as the Greek. [. . .]

I would like to see more research. Anyone who goes to an MLA Convention, which I do every year, finds all those people doing the most obscure research on Henry James and Joseph Conrad and so forth, while there are any number of other topics that no one ever looks at. I think Constance Garnett performed a terrific service for English and American literature, and that is worthy of a thoroughgoing study, to see what happened in the translations. And I don't think it has to be primarily evaluative, putting them down, claiming that you can't read them anymore, but simply to see what happens in those translations. We need a methodology, which we really don't yet have for analyzing such translations, describing how they came out this way, and further accounting for their impact over such a long period. It seems to me there is fruit here for dozens, hundreds, of dissertations that could be much more interesting than many of the dissertations that people are working on now, because they're plowing so many fields that have been plowed countless times before.

Comment: Very often the problem is that of convincing professors of literature of the seriousness of a particular line of research and dissertation topic. What can we do?

A: Yes, that was my point at the beginning. It seems to me that they misunderstand how the history of literature works . . . I don't know. Discussions like this are one way.

SYBREN POLET

Four Poems

Translated from the Dutch by James S. Holmes

Sybren Polet was born in Kampen, the Netherlands, in 1924. Polet is numbered among the most important poets of his generation. He has travelled extensively and lived for long periods is Sweden, Spain and the United States. He has published translations of modern Swedish poetry besides an anthology of modern world poetry and two very early anthologies of science fiction. His poetry, in fact, derives some of its major themes from the sphere of science fiction—those of metamorphosis and rebirth. One of his books of poetry, *Town of Rebirth* (1960), bears Paul Eluard's motto: 'Our birth is perpetual', and another, *Person/Unperson* (1971), Virginia Woolf's 'I am made and remade continually. Different people draw different words from me.' These themes are also connected to a notion of social consciousness, a notion of man-as-a-social-organism, present since his second volume, *Orgnanon* (1958) and becoming more explicit in *Metaconcrete Poetry* (1962): a fellow-man as another I. The techniques of objectivization and many transformations in these poems often partake of modern technology, in an attempt to democratize the difference between poetic and apoetic spheres of language. Later his work focuses more obviously on the social motives already present in his earlier work —the volumes *Person/Unperson, Illusion & Illumination* (1976)—and on language as such, *Language Figurations I and II*

(1983). Apart from his poetry, Polet is also known for his prose, essays, plays, fairy tales, which makes him one of the most interesting all-round writers in Holland. (*James S. Holmes*)

Lady Godiva on a Scooter

Ride, ride,
on your scooter of flowers,
your prime mover
that knows no blocks.

Adjust your horn
to a frequency so high
that the prying windows shatter
and the faces blush like shame-red roses.

Ride, ride so close to them
the cops start writing poems
that they read to their wives at night
in motorized voices
and the university librarian, irritated,
takes off his hat
to mop literature from his brow.

Ride then, this is a land overflowing
with milk and eros, where even
the simplest images bathe in cucumber milk
and kings have no care for wars and growing older.

Ride then, ride, on your man-shaped colt
that does not thirst after distances
unattainable,
only after the hand of its recreator
that regulates its temperament,

restrains its progressive dreams,
trims its cosmic mane.

The wind that springs up from a flower
becomes an artistic flower in your hair,
the trees—old jesuits and parsons—bowing trees
whose leafage kisses your breasts;
the sun's excited finger counts your vertebrae one by one,
and if a hand should flutter against your leg
don't think it's a falling leaf,
it is a court poet's approving hand.

Ride then, ride; this land has no bounds,
no gods as customs officials, eyes
with no one behind them.
No architecture, cool, constrains you, no
flora, tame or wild, controls you,
not even a flaming word bars the way with a sword.

Ride then, ride, ride—for we are
the poor in body
of whom it is sung
that they shall inherit the earth.

Day's Work

I carry my shoulders like two black owls:
they have no strong wings and no darkness to make them
  fly,
only perhaps the thought that like me is weighed down by
  this politicized daylight.
Ah, I won't give darkness for nothing, I walk away
and it's a proud soldier who can detect me with his detector.

This then is my day's work:
I help my thousands of invalids hide, my
refugees flee. My mystics

I sympathize with in the spirit: this darkness is sufficient
onto them.
If I'm asked: How can such a thin voice give shelter
to so many and such lifesize invalids? I answer: Living is
good for them,
they beg not neither do they drool, my tongue is medicinal
and their faith is as clear as heredity.

Thus I pass their day, according to talents and lights
apportion the available tasks, among which:
wearing a moustache and smiling, planting words, being a
soldier,
among which: the giving and taking of hands, and,
beneath a ruddier light, the listening to
armies of footstamping ideas and fetuses.

But at night . . . at night I rage at dictators,
dictators and their holloweyed music
that stands up straddle-legged as their uniforms rest.
I sit in their darkness as in a definitive home.
I tie my dog to the post of their thoughts.
He bites not or only nips, but growls, lions
and swells beneath their blustering breath.

Well, growl then, swell, growl, I evade
your thoughts as a poet,
or the taxman as a pen name.

Enemy

My enemy is my shepherd,
my shepherd shall not want,
he lies down in my voice like a bleating lamb.
Theoretically viewed there is no lack in him.
On his two apartment buildings he stumbles through my
province,

and his head runs over like a village of kine and passion.
When he speaks, his voice defaces the freefloating voices
and the sound is of broken spectacle glass
which is the stuff that concrete dreams are made of.

O, he lacks not of eyes and fine matter,
his love life encompasses an entire scientific order
and he populates equally air and earth,
just as two images silently propagate themselves in a
poem—
In this way he causes the mightiest politicians to kneel.

Warm are his arms of wrath and I rest in them
like a motor in an organizer, like a motor
I know no better, know nothing. His reality
is my reality and therein is no deceit
possible: I am one of the images that propagate him.

Heartening are his shoulders of pain, his hair of pain and
grief.
So violent are his punishments that I explode with
laughter into sand
and at night I voluntarily help to murder
a year-old statue.
For oh, to deny him too is of no avail; for he
does not consist of projections, nor does he thrive upon
identifications,
my enemy is more real than a paradox, more fullblooded
than a bull of an antithesis.

And this is how he triumphs as my mouth denies him:
he dissolves himself in my enemy and my enemies dissolve
into him.

This is my enemy.
My enemy shall not want.
He lies down in my voice like a leech.

Democracy

1

Not another reality, but the same reality
otherwise. Not otherwise, but more visible,
translucent. I have built a city.

I have built not a city but a poem.
Poems are clearer than cities
and words are stronger than the houses of a city.
I have built a city.

Houses are the thoughts of a city.
People are to hide in them
    like a flapping idea in a bird.
Human houses tip over easily.

Since logic is relief and intelligence a hand
—infinite extension of the body, servoman
to helpless and mechanical man—
I too am man's servoman,
whoever he is and whencever: eclectic, electric, desperate,
  mild.
I am my own servoman.

Because, and this is not the uninhabitable world of the air,
this is the inhabited world of poetry,
and all that makes sense finds its sense in poetry.

That's my one reason for writing poetry.

Or counter-poetry.

2

That's my one reason for writing poetry,
while I call the word by its (double) name.
I have a double name.

I have a chronic name:
one for the city, one
blank, with which I can laugh, drink, laugh.

An urbanized Indian
smokes away his despair with an electronic cigaret;
he grows wise as a cyclop, i.e. he knows
  how wise a cyclop is; I,
I don't.

And the fakir who phosphorizing regulates
from his midstreet island the spiritual traffic in his city
wants to be more than his dream;
he is a skeleton of athletic thoughts:
Atlas.—This poet too is an athlete: he carries
a head on his shoulders.—Atlas? At every thought
he grows older; well, I,
I don't. Because:

No system is worth its spirit,
but the spirit is worth a system;
the city is worth a structure
and a robot its poetry.

That's my democracy.

3

In uranium's book
I am an invisible radio.

I sit in my radio
like an Indian in his teepee.

The sun is my impresario,
my god is a sunlamp.

I am my own electrician,
I am my own continuum.

Living chiefly on spiritual helium,
1000 Volts is my maximum

tension. I-contact. I
annihilate myself, originate,

an intelligent protozoon in
initial chemical refinement, later

a fish of will: I lie out of water,
in a coma, subject to your electrolysis;

sneezing I awake, I see, see
you; I smell your tetrachloride breath.

– Thank you Ma'am, I'm feeling better.

Then I stand upright, monkey, I, nipplewhite zombie,
warrior ant, art ant, eyes front and ever onwards.

Content with even $(\frac{1}{8})^2$ of life
I versify myself slowly into life.

The Spirit has yielded the spirit,
The Word has given its word.

KITTY M. VAN LEUVEN-ZWART

Building Bridges: James S. Holmes and Translation Studies

What particularly characterizes translation studies, so called, is the gap between theory and practice. This gap manifests itself in two ways: translation theories often lack any connection with the everyday experience of practising translators, while translators generally ignore or belittle the efforts of theoreticians in the field. Thus, the opening sentence of a recent Dutch publication on translation reads: ' . . . the mere mention of the discipline [of translation studies] among translators is enough to provoke jeers and laughter' (Langeveld 1986:11; see also van Leuven-Zwart 1988: 701). Needless to say, if translation theories were not, for the greater part, mere products of ' . . . armchair rumination in splendid isolation . . . ' (Holmes 1988: 102), Langeveld's observation would not apply.

One of the most fervent advocates of a rapprochement and, to my knowledge, among the first whose proposals in this respect were of a fundamental, metatheoretical character, was James S. Holmes (1923–1986). Two years after his death, Raymond van den Broeck and Hans van Marle have collected and edited Holmes' scattered papers in the volume *Translated! Papers on Literary Translation and Translation Studies* (Holmes 1988).

What strikes me now, in reading *Translated!*, is firstly how the papers reflect Holmes' personality, and secondly the consistency and coherence they show, notwithstanding the fact that they were written over a period of more than fifteen years. In discussing these qualities, I will try to elucidate the importance of Holmes' role in the field of translation studies.

Everyone who has known Jim Holmes personally will agree with me when I say that one of his most distinguishing and endearing features was his tolerant and relativizing attitude towards life. This same attitude manifests itself in his papers, especially in the volume's first part: 'The Poem Translated.' Here, his own experience as a translator explicitly constitutes the basis for his theoretical observations, dissolving the division between the theory and practice of translation in a natural and elegant way.

Holmes' tolerant and relativizing attitude becomes particularly clear in his concept of 'metapoem', elaborated in the first two articles, 'Poem and Metapoem: Poetry from Dutch to English' and 'Forms of Verse Translation and the Translation of Verse Form.' In his view, a translation of a poem should be considered a comment, an interpretation of that poem:

> ... the verse translator, by virtue of the choices he is required to make in his pursuit of the illusion of unity, presents one possible interpretation (out of many) of the original poem, re-emphasizing certain aspects at the cost of others (Holmes 1988: 51).

What is implied is the rejection of the notion of an 'ideal', 'ultimate' or 'absolute' translation which eliminates all others:

> It is for this reason that there will always be need of more than one translation of any poem of importance, since several translations present more facets of the original than any one can do (idem: 51).

Typical of Holmes is the fact that he applies this relativizing view on translation not only to the work of others, but to his own translations as well. After having discussed, for example, the problems he was faced with when translating Nijhoff's poem 'De Grot', he states:

Some day another translator (or even I) may work out another version of 'De Grot' that does justice to its formal structure. If he succeeds in this without inflicting greater damage upon other aspects of the poem than I have done, his translation will deserve to supplant mine. More probably, the two will continue to exist side by side, each supplementing the other as a somewhat differently faceted commentary on the many-faceted original (idem: 58).

Also implied in Holmes' view on translation is a non-normative, non-prescriptive posture in regard to his own and other people's translations. His aim is to show which choices the translator had made and how these choices compel the translator to emphasize certain aspects of the original text, while 'underexposing' or giving up others. It is this descriptive approach which guards him from the 'vain complacency' (Van den Broeck in Holmes 1988: 2) we usually are faced with when translators comment on their own or other people's translations.

Holmes' tolerant and relativizing attitude is reflected not only in his view on translation and translations, but manifests itself also in his language. His articles show that scholarly papers on translation need not be tedious or unreadable, and that scientific need not be synonymous with jargon-ridden and arid. On the contrary, to read Holmes is almost to hear him talk again, as if one were attending his lectures. His is what may be called a genuinely elegant style, clear and simple, understated and ironic. It is the style of a man who has succeeded in reconciling both sides of a dual personality: the artist with a passion for poetry and translation, and the scholar who enjoyed theoretical inquiry and reflection on these subjects.

As was stated earlier, when I read the volume I was struck not only by the way in which the collected papers bear witness to Holmes' personality, but also by their consistency and coherence. Each of these articles stands as a testimony to Holmes' efforts to secure a basis for the development of translation studies into a mature and independent academic discipline. In Holmes' view, apart from the gap dividing the theory and the practice of translation, two other factors account for the absence of such a basis: the rift and rivalry between the linguistic and the literary approach to translation research, and the dispute over whether translation phenomena ought to be investigated from a normative or from a descriptive standpoint.

When, in 1968, Holmes started to publish in the field of translation studies, the discipline was dominated by the linguistic school. This particular approach, flourishing especially in Germany where it was headed by the so-called Leipziger Uebersetzungswissenschaftliche Schule, originated in the post-war optimism about the potential of machine translation. Translation was seen as the recording of messages, as an eminently linguistic operation which should be investigated with methods comparable to those applied in the natural sciences. Thus, there was no room for matters not directly pertaining to the phenomenon of language, such as the situation in which language is used, the intention of speaker or writer, the effects of utterances on hearers or readers, or the communicative functions texts may perform.

As a consequence, literary translation was regarded as outside the field of 'translation science' or, as the Germans call it, 'Uebersetzungswissenschaft'. This conception of the field, the 'freie Spiel Sprachgestalterischer Kraefte' (Wilss 1977: 21) excluded such vague matters as intuition, talent and creativity as being non-amenable to objective, scientific investigation. Thus, in post-war translation research, two distinct approaches emerge, a linguistic and a literary one, literary translation research substituting 'all too frequently . . . impressionism for methodology' (Holmes 1988: 100), while linguistic research claimed to be the one and only genuinely scientific way of studying the phenomenon of translation.

Another split occurred in view of the fact that linguistic

publications on translation theory were very hard to read, since in their pursuit of scientific objectivity, linguistically oriented translation scholars used a highly formalized language and peppered their papers with intricate formulas and diagrams. Thus, these papers were accessible to specialists only, failing to induce translators to become acquainted with translation theory. Helping to exacerbate was the fact that most of the translation theories developed within the linguistic tradition were the result of what Holmes, as previously quoted, calls 'armchair rumination in splendid isolation,' deductive theories which lacked any relationship with the everyday experience of translators and with the products of their activity, the translations themselves:

> Many of the weaknesses and naivetés of contemporary translation theories are a result of the fact that the theories were, by and large, developed deductively, without recourse to actual translated texts-in-function, or at best to a very restricted corpus introduced for illustration rather than for verification or falsification (idem: 101).

In Holmes' view, neither the linguistic nor the literary orientation in translation research could provide a sound basis for the development of translation studies into an autonomous, mature discipline so long as they approached translation exclusively from a prescriptive, normative angle, excluding the descriptive and non-normative. Literary translation scholars 'too often erred in mistaking their personal, national, or period norms for general translation laws' (idem: 99), while theories developed within the linguistic tradition failed to explore and explain the translation phenomenon, as might be expected, but rather provided rules and prescriptions for the production of a good translation:

> It seems to me, in fact, that many of the theories of translation that we have had up to now, while pretending to be theories of the translation process, are in fact theories for translation didactics. They are giving us material to train translators . . . It seems to me that before we can know how to train translators, we have to know what takes place in

the translation process and we have to know what translated texts are (idem: 95, 96).

Holmes set himself to bridge the gaps described here. In the first part of *Translated!*, 'The Poem Translated', an impressionism prevalent in literary translation complements a systematic, methodological approach, comparable to the manner of linguistic translation research. Thus, Holmes shows that the linguistic and the literary approaches to translation do not exclude one another, and that even such vague and 'unscientific' matters as creativity and intuition, which in the eyes of linguistically oriented translation scholars exclude a 'scientific' treatment of literary translation, may be investigated systematically.

At the same time, the papers in 'The Poem Translated' show how fruitful and illuminating the inductive and non-normative approach to translation may be. The inductive view, with its constant recourse to the practice of translation, guards 'the scholar from sterile theorization' (van den Broeck in Holmes 1988: 2), while the descriptive, non-normative approach guards 'the translator from vain complacency' (idem: 2). Thus, the first part of the volume may be regarded as the actualization of Holmes' view on translation studies: a view which comprises literary and linguistic, inductive and deductive, descriptive and prescriptive, a view, in short, which secures the fruitful exploration of the theory and practice of all translation phenomena.

It is in part two, 'Studying Translation and Translation Studies', that Holmes seeks to establish a basis for his ideas. In the first paper, 'The Name and Nature of Translation Studies', he discusses the objectives of the discipline. In the other papers in this part he elaborates and clarifies the views set down in the first. I will concentrate now on those ideas which, in my opinion, are still of the utmost importance for the development of translation studies, but which, surprisingly enough, have so far not met with sufficient response.

In 'The Name and Nature of Translation Studies' Holmes makes use of Francis Bacon's distinction between research which is 'of light' and research which is 'of use'. Here, 'of light' refers to pure or theoretical research in translation studies, research the objective of which is to describe and explain the phenomena of translation:

As a field of pure research—that is to say, research pursued for its own sake, quite apart from any direct practical application outside its own terrain—translation studies . . . has two main objectives: (1) to describe the phenomena of translating and translation(s) as they manifest themselves in the world of our experience, and (2) to establish general principles by means of which these phenomena can be explained and predicted. The two branches of pure translation studies concerning themselves with these objectives can be designated *descriptive translation studies* or *translation description* and *theoretical translation studies* or *translation theory* (idem: 71).

In the field of descriptive translation studies Holmes distinguishes three areas. In the first area, the translation process, what takes place in the translator's mind, constitutes the object of research. In the second area, the focus is on the product of this process: the characteristics of existing translation. Research here may take the form of a comparison of translation and original. The aim of this type of research is to identify the translators' interpretations of original texts and the norms and strategies applied in different periods and in different cultures. In the third area, the role played by translations in the recipient culture is the object of investigation. All three areas are relevant to both literary and non-literary translation.

As to theoretical translation studies, Holmes distinguishes six areas or types: medium-restricted, area-restricted, rank-restricted, text-type-restricted, time-restricted and problem-restricted. In 'Translation Theory, Translation Theories, Translation Studies and the Translator', however, he combines descriptive and theoretical translation studies:

No one, at least to my knowledge, at least in the West, has yet even dared to face up to the need for a 'field theory' of the entire range of the translation phenomenon. Such a field theory would be built up of a number of partial theories .... We need a theory of the translation *process*, that is, the theory of what happens when people decide to translate something. We need a theory of the translation *product*, that is to say, what is specific to the translated text as a text; in what ways is it similar to and in what ways is it different from other kinds of texts, literary or other. We need a theory of the translation *function* that is, how the translation works in the recipient society (idem: 95).

By combining the three areas which he distinguishes in descriptive translation studies with those of theoretical translation studies, Holmes takes a significant step forward. Insisting on the interdependence of these two branches, he establishes an empirical, inductive basis for the theory of translation.

Translation research which is 'of use' constitutes for Holmes what is termed applied translation studies. The aim of this kind of research is to develop the means and methods which are of direct use to the translator. Again, Holmes distinguishes a number of areas. In one area the researcher's concern is with translation dictionaries, translation grammars and terminology banks, whether or not computer-linked. Two other important areas are translation criticism and translation didactics. In translation criticism the aim is to develop methods by which translations can be more adequately evaluated than has been the case till now. Research in the area of translation didactics looks for ways in which teaching can be made more systematic and more effective.

In 'The Name and Nature of Translation Studies' and the other papers in the second part, Holmes provides the means of healing the divisions among translators and translation theorists, such as that which exists between the linguistic and the literary approaches to translation studies. Holmes' view is that the discipline cannot grow to maturity and independence if it excludes categorically and systematically one or other text-type or approach. Both literary and non-literary translation belong to the field of translation studies; the literary and the linguistic should not exclude each other, but should interact as productively as possible. At the same time, Holmes underscores the interdisciplinary character of the discipline: not only linguistic and literary research, but also, text-linguistic, psycholinguistic, sociolinguistic and semiotic research can legitimately be employed in investigating translation phenomena.

There is another division with regards to the role of translation studies: Is it to be prescriptive and normative, or is it to be descriptive and non-normative? Holmes accommodates both views. The descriptive approach is by its nature connected with pure translation research, whereas the prescriptive, normative approach is best suited to the demands of applied translation studies:

> The first three of these partial theories, [i.e. the theories of the translation process, product and function; see above] I feel, should be non-normative. They should be attempts to provide models by which we could analyze existing situations, describing not what the situation should be but what it is. The fourth, on the other hand, the theory of translation didactics, is necessarily normative. We have to make decisions about how to train translators whether we know the answers or not, simply because translators have to be trained (idem: 95, 96).

In Holmes' view, both pure and applied translation studies can only come to full bloom if they feed each other continually:

> . . . descriptive, theoretical and applied translation studies have been presented as three fairly distinct branches of the entire discipline, and the order of presentation might be taken to suggest that their import for one another is unidirectional, translation description supplying the basic data upon which translation theory is to be built, and the two of them providing the scholarly findings which are to be put to use in applied translation studies. In reality, of course, the relation is a dialectical one, with each of the three branches supplying materials for the other two, and making use of the findings which they in turn

> provide it . . . In view of this dialectical relationship, it follows that . . . attention to all three branches is required if the discipline is to grow and flourish (idem: 78–79).

The last gap bridged by Holmes is that which in my view, largely characterizes the field of translation studies, thoroughly blocking its development, namely between the theory and practice of translation, between theorists and practitioners. Holmes contends that researchers, both in pure and in applied translation studies, have failed to forge links with the practice of translation. The danger, Holmes claims, is that eventually this failure will dam up research. Translation, of course, will go on as usual. It has managed without translation research for centuries and will not suddenly come to a halt for the lack of it. Nevertheless, there is no denying the fact that the practice of translation *can* benefit from research. I will conclude with a quotation from the paper 'Translation Theory, Translation Theories, Translation Studies and the Translator,' which eloquently demonstrates that there is reason enough for practising translators, too, to attempt a reconciliation.

> Reading translation theory and thinking about translation problems, one reaches a dead point when one feels blocked in every direction and can't do anything; such is the frustration about the whole situation that one thinks about giving up translation. At that point, translation theory can do a great deal to help make us aware of a great many more things to help us over that dead point so that, from that point on, we could translate much more consciously, being much more aware of the choices. We can get rid of some of the almost childlike beliefs that we have worked with for a long time (it has to be done this way, this is always wrong, this way is always right) and realize that in this situation, *this* is perhaps the most appropriate tactic and in that situation, *that* is perhaps the most appropriate tactic. If translation theory, even at its present state, can give us some more awareness of what we are doing as translators and help us to think and become conscious of our activity, then I think it has fulfilled and important role (idem: 97/98).

REFERENCES

Holmes, James S., *Translated! Papers on Literary Translation and Translation Studies*. Amsterdam: Rodopi, 1988.

Langeveld, Arthur, *Vertalen wat er staat*. Amsterdam: Uitgeverij de Arbeiderspers, 1986.

Leuven-Zwart, Kitty M. van, 'Vertalen en vertaalwetenschap: tweespalt of eenheid'. *De Gids*, honderdeenenvijftigste jaargang, nr. 9, 701–709, 1988.

Wilss, Wolfram, *Uebersetzungswissenschaft, Probleme und methoden*. Stuttgart: Klett, 1977.

THERESA KI-JA KIM

The Korean Shaman Song of Creation

INTRODUCTION

The Shaman Song of Creation recounts the birth of the cosmic universe from Chaos/Darkness/Void.[1] It establishes the origin, structure, and operation of the universe and marks the distinction between that which is sacred and that which is profane. These notes introduce the opening of the Song in English translation. The term *shaman* is derived from the Tungus *saman*, which originated from the Sanskrit *śramaṇa* (Eliade 1964: 495), meaning 'the one who practices asceticism' (Blacker 1975: 318). The term describes a priest or priestess of the Arctic Siberian and Ural-Altaic peoples of Central Asia whose religious duties were: (1) to link this world to that of the unseen; (2) to induce a

[1] Chaos/Darkness/Void is a loose translation of two Chinese characters, *hong-don* and *am-húk* which literally mean 'intermingled confusion' and 'darkness.' Due to the inclusive nature of Chinese characters, these translations cannot be interpreted literally. There is no English equivalent for these terms which together correspond to the meaning implied by Chaos/Darkness/Void. Since all things are born of Chaos/Darkness/Void, Chaos is not chaos in a literal sense, but rather an intermingled state of the essence of all things. Darkness cannot be interpreted literally either, since the sun has not yet been born. Void should not be understood as emptiness, but rather as formlessness which contains the potential for all forms. Chaos/Darkness/Void is, therefore, an expression of an all-inclusive state of everything/nothingness, best described by the dialectic relationship of existence/non-existence in perfect equilibrium.

state of trance-possession in which the soul would leave the body to take a cosmic journey upward, downward, or into the horizontal regions; (3) to call on the appropriate spirits to manifest their intentions; and (4) to lead the tribe, counsel the people, make divinations and prophecies, perform miracles, cure diseases, reverse bad fortunes, prolong good fortunes, change the identity of things, console dead souls, and control natural and supernatural phenomena.[2] Shamanism is regarded as the oldest religious phenomenon. Nowadays the term is used loosely for any mystic magico-religious cult that shares some of the original beliefs and practices of the Arctic and Ural-Altaic prototype, wherever it is found.

Koreans are believed to be the descendants of Ural-Altaic peoples (Park, 1963: 23–4). The indigenous religious phenomenon of Korea is shamanism. Korean shamans' central tenet is that each element in the universe is governed by its own force, which has both good and bad tendencies. These forces have concrete names, such as The One, The Great One, and The Shining One, as well as more abstract ones like Spirit, God, and Deity (here the term *deity* has been used, in preference to *god* with its Judeo-Christian associations). Each deity is placated by shamans through rituals so that the beneficial element is activated for the good of humans.

The origins of Korean shamanism have been traced back to the ancient nomad Turks of the north, as well as to southern Asia. Apart from some fragmentary records of shamanic practices as early as the three Korean Han (*Chinhan, Pyonhan,* and *Mahan*) periods around 1500 BC, there is no written account of shamanic ideology (Eliade, 1964: 461–2).

The oldest written account of the history of the Korean people is *Wei-shu* (The Book of Wei), which was compiled by the

[2] The definition is loosely based on 'Shamanism', *Encyclopedia Britannica, Vol. XX,* 1973; Eliade, 1964, 454; Yu, 1975, 63.

Chinese scholar Ch'in-shi (233–93 AD) during the Chin dynasty. The *Wei-shu*, one of the sixty-five volumes of *Sankuo-shi* (*The History of the Three* [Chinese] *Kingdoms*), contains a 'Section on the Eastern Barbarians.' This section documents various 'barbarians who inhabited the area north-east of China known as Manchuria, the Korean peninsula, and the Japanese islands (Illyŏn [1280]; 1972.)

The *Wei-shu* deals with the first Korean nation, *Chosŏn* (2333 BC), literally meaning the 'Kingdom of Morning Splendor.' The birth myth of the founder *Tan'gun Wanggŏm* (The King of Sandalwood) from the marriage between *Hwan-ung Ch'ŏn-wang* (The Shining Male Heavenly King) and *Ungnyõ* (She-Bear) exhibits the shamanic motif of initiation, transformation, and marriage between the super-natural and a mortal. Some scholars 'believed that *Tan'gun* was simply a great Shaman, and that his worship on the great altar on Kanghwa was simply a shamanistic performance' (Clark 1961: 176).

The birth myths of the three following kingdoms of *Koguryŏ* (37 BC–668 AD), *Paekje* (28 BC–660 AD), and *Silla* (57 BC–935 AD) contain the same motifs and provide some evidence of oracle-prophecy by founding kings and queens. In the *Silla* dynasty, the name of the second king, *Ch'a-ch'a-ung* (ruled 4–25 AD), means 'the male shaman.' Court annals of *Silla* also provide abundant accounts of Queen *Sŏndŏk* (ruled 632–646 AD), who made prophesies and predictions of the future by reading natural signs (Illyŏn [1280]; 1972, 52; 73–5). The forty-ninth King, *Hŏn'gang*, often entered a state of possession and sometimes danced like the South Mountain Deity that he claimed to envision. The invention of the Old Man Deity dance is credited to him, and it is believed that he had his craftsmen carve a mask in the likeness of the deity he envisioned (Lee, 1969: 29).

The once dominant shamans became progressively demoralized with the advent of such foreign-born religions as Buddhism, introduced in Korea in 372 AD (Clark, 1961: 27), Taoism, and

Confucianism. Korean shamanism itself has been modified by the spirit worship of Taoism and has borrowed extensively from Buddhism. Yet shamanistic beliefs have survived in the very consciousness of the Korean people.

There is no written record left as to how shamanistic rituals were to be performed. Shamans learned the songs through apprenticeship to master shamans. No individual shaman is responsible for the composition of songs which have been assembled over many centuries of oral transmission, doubtless undergoing changes in this long process. The songs contain lyric poetry, though not necessarily following any rigid forms. There are also narratives, recitatives, and chants. Refrain is a common device. Songs can be awe-inspiring, grand, urbane, solemn. Some are chanted, while others are more melodious. Since the songs are accompanied by musical instruments and sometimes accommodate dancing, there are rich rhythmic patterns.

The origins of the Shaman Song of Creation or *Ch'ŏnji Kaebyŏk-ga* or *Ch'angse-ga* cannot be determined. *Ch'ŏnji* combines two Chinese characters, *ch'on* (heaven) and *ji* (earth), connoting the universe. *Kaebyŏk-ga* consists of three words: *kae* (opening), *byŏk* (the primeval beginning), and *ga* (song). Thus, the literal translation of *Ch'ŏnji Kaebyŏk-ga* is The Song of the Great Beginning of the Universe. *Ch'angse-ga* consists of three Chinese characters: *ch'ang* (creation), *se* (universe), and *ga* (song). *Ch'angse-ga* is, therefore, The Song of the Creation of the Universe.

Fragmented versions of The Shaman Song of Creation are in evidence throughout Korea. In North and South Hamgyŏng Provinces, as well as in the northeastern-most part of Korea, the song is called *Changse-ga* (Kim, 1981: 167). The song is also sung in the *Yongwang-gut* or Rite of the Cauldron Deity, in the Puyŏ area of South Ch'ungch'ŏng Province, where the southwest coastal region adjoins the Yellow Sea (Kim, 1971: 105). Further south, in the port city of South Chŏlla Province, in the region of Kunsan, the song is called *Chiduso* (Kim, 1976: 81). The best preserved version, however, is *Ch'ŏnjiwang ponp'uri* of Cheju Island.[3]

The shamanic hymn of Cheju Island is called *Ch'ŏnjiwang ponp'uri. Ch'ŏnjiwang* derives from three Chinese characters, *Ch'ŏn* and *ji* (heaven and earth), and *wang* (emperor, king, or great one). *Ponp'uri* is a composite of two words: *pon* (the origin), and *p'uri* (the disclosure, revelation, or unfolding). Thus, *Ch'ŏnjiwang ponp'uri* translates as The Unfolding of the Origin of the Deities of Heaven and Earth.

This mythic hymn is sung in the opening episode, *Ch'ogamje,* of the Great Ritual. The term *Ch'ogamje* is composed of three Chinese characters: the prefix *ch'o* (primordial beginning), *gam* (the great one), and the suffix *je* (rite or ceremony). As its title indicates, the *Ch'ogamje* celebrates the beginning of the universe. The mythic hymn itself is frequently referred to by the same name.

After various stages of purification, the first episode of The Great Ritual takes place. The shaman stands in front of the offering table, holding a fan and handbells. The clients stand nearby, rubbing their hands together and bowing from time to time in a reverent manner. The shaman then states the reason for holding the ritual. She utters the name of each client, each client's date of birth and address, and the time and place in which the ritual is taking place:

Hear ye,
Hear ye,

One and all,
One and all,

[3] An island sixty miles south of the Korean mainland.

The Rite of [name of the deity]
Is about to be held;

In this auspicious year,
The year of the [dog, dragon, etc.];[4]

In this auspicious month,
The month of the [dog, dragon, etc.];

On this auspicious day,
The day of the [dog, dragon, etc.];

This ceremony we hereby
Do hold.

The shaman now seats herself at the offering table and begins to sing the Shaman Song of Creation while playing an hourglass-shaped drum.[5] The origin of the universe is unfolded in this way (sung by the male shaman, Chu-byong Chong, translated by this author):

The Shaman Song of Creation

In the great beginning
There was void.

Heaven and
Earth,

With no separation
Between them,

Were in the immense
Darkness of void.

[4] Name of the year, according to the twelve earthly zodiacal signs (see Commentary) which should be in accordance with the time and date the performance is held.

[5] In the shamanic profession in Korea, females outnumber males ten to one.

From within
The void

Emerged motions,
The birth of the world.

First, Heaven appeared,
Head first,

At the hour of *kab-ja*
The first hour,

On the day of *kab-ja*,
The first day,

In the month of *kab-ja*,
The first month,

In the year of *kab-ja*,
The first year,

In the direction of *ja*,
The rat.

Second, Earth appeared,
Head first,

At the hour of *úl-juk*,
The second hour,

On the day of *úl-juk*,
the second day,

In the month of *úl-juk*,
The second month,

In the year of *úl-juk*,
The second year,

In the direction of *juk*
The bull.

Third, Man appeared,
Head first

At the hour of *pyŏng*-in,
The third hour,

On the day of *pyŏng-in*,
The third day,

In the month of *pyŏng-in*,
The third month.

In the year of *pyŏng-in*,
The third year,

In the direction of *in*,
The tiger.

Fourth, the deities appeared,
Head first,

In the direction of *kŏn*,
The heavens.

Fifth, Multitudes of creatures appeared,
Head first

In the direction of *kŏn*,
The earth.

The line of separation appeared
Between heaven and earth.

The division became
Greater and greater.

The mountains high and low
Appeared on earth.

And the rivers
Began to flow.

The distinction between
Heaven and earth

Became clearer
and clearer.

From heaven,
The blue dew descended.

From earth,
The black water sprang.

They mingled together,
as *yin* and *yang*

And all things in the universe
Were created.

First created were
The stars.

The Young Cowherd Star
In the east,

The Weaving Maiden Star
In the west,

The Old Man Star
In the south,

The Great Northern Seven Stars
In the north,

And in the Center was
The Three Great Stars.

Many lesser stars
In the sky

Were born to occupy
Their places.

Darkness was
Still flowing.

Only the clouds were
Yet to come.

In the west,
The white cloud,

In the south,
The red cloud,

In the center,
The yellow cloud,

In the north,
The black cloud,

And in the east
The blue cloud.

The rooster of Heaven
Set his neck high,

The rooster of the Earth
Flapped his wings,

The rooster of Man
Beat his tail feathers together.

As they crowed all together
In a roaring sound

The dawn
Broke through

In the direction of
Kab and *ul* of the east.

Two suns and
Two moons appeared

In the sky
By the Heavenly Emperor

The King of Heaven
And Earth.

Heaven and earth
Appeared brightly.

Still there was
No order,

With two suns and two moons
In the sky.

Many thousands of men
Died of heat during the day,

Many thousands of men
Died of cold during the night.

All the creatures,
On Earth,

The trees and
The grass,

The birds and
The animals,

All spoke
In their own languages.

There was
No distinction

Between all the things
Created.

Even between men and
The deities.

When men were
Called upon,

The deities answered
Instead.

When spirits were
Called upon,

Men answered
Instead.

(Hyŏn, 1976: 11–12)

From here on, the myth revolves around the Heavenly Emperor. To bring order into the chaotic world, the Heavenly Emperor descends to earth and institutes laws. Later, he marries the earth queen. Returning to heaven after announcing to the earth queen that she will give birth to two sons, the emperor instructs her to name the elder son, *Taebyŏlwang*, or Big Star King, and the younger son, *Sobyŏlwang*, or Small Star King. The two sons are to be raised to govern the world and so must compete with each other to decide who will rule the upper and who the under world. The younger brother outwits the elder and begins to govern the upper world, while the older brother is given jurisdiction over the under world. The younger brother, however, fails to govern wisely and great disorder results. Eventually, he permits the older brother to rule in his place.

The older brother shoots down one of the two suns and one of the two moons, thereby preventing the life forms from dying of heat or cold. He reserves spoken language for men and deities only by spraying glue to seal the mouths of all the other creatures. However, he fails to complete the task of establishing order in the world, so that murder, theft, sickness, ill-will and all the other evils continue (Hyŏn 1976: 11–20).

The Song of Creation ends here, thus concluding the first

episode, *Ch'ogamje* of The Great Ritual, each of the remaining episodes being dedicated to a specific deity. The ritual concludes with the Rite of Burning.

COMMENTARY

The first stanza of The Shaman Song of Creation envisions the world emerging from Chaos/Darkness/Void. As with song itself, there is no way of determining when and how this notion came into being. Chaos/Darkness/Void is an all-inclusive emptiness containing all things in their embryonic and neutral state. To Korean shamans the idea of the Void is neither pure nor impure. The Void is viewed simply as existent and is considered sacred on account of its eternal all-inclusiveness, non-separateness, nothing/everything emptiness. This state is Almighty and Absolute.

The 'impersonal' nature of this void suggests that historically the myth even predates the idea of man as superior to all things in the universe, which led to the creation of a deified figure, or God in the form of man.

The second stanza sings of the emergence of 'motions' from within Chaos/Darkness/Void. The equilibrium of the void is disturbed and a universe is born. The all-inclusive potential neutrality is broken, as essences begin to interact to form matter. The universe is not created by a force outside the Void; rather, it is triggered by vibrations from within. The Korean shaman's vision of the birth of the universe is systematic and concrete: 'First, heaven appeared, head first' and 'Second, earth appeared, head first' and so on. Thus, time and space, heaven, earth, and man together form the universe.

There are altogether ten symbols of 'heaven' and twelve zodiacal symbols of 'earth', which correspond to *kap, úl, pyŏng, chŏng, mu, ki, kyŏng, sin, im, kye,* and the rat, the bull, the tiger, the hare, the dragon, the snake, the horse, the lamb, the monkey, the cock, the dog, and the hog. This combination of ten heavenly signs and twelve earthly signs forms a sixty-year cycle, which when completed, begins anew. In Korea, the sixtieth birthday, called *hoe-gab*, or 'return to the first,' is cause for great celebration.

The birth of cosmic time and space begins the cycle of the birth and death of all things. Whatever is born must die; nothing is eternal. Like the seasons, everything is constantly changing, endlessly repeating the cycle of birth-growth-death. Not only humans, but the entire universe must one day return to Chaos/Darkness/Void, thus completing the cycle. A pattern of return to the Void is established for all things in the universe. Korean shamans consider the cosmos profane, because of its transient nature and its adherence to the cosmic rule of change. However, all entities in the universe are potentially sacred, since it was born of Chaos/Darkness/Void. Everything in the universe is linked, through its core, to the macrocosmic Void. All entities are therefore simultaneously sacred and profane as manifested in the deities possessing both benevolent and malevolent tendencies.

In the Korean shamanic view, death represents a regression to the original neutrality. When matter is dissolved by death, it no longer possesses time as it ceases to occupy space. The regression of each individual entity to Chaos/Darkness/Void is a microcosmic reflection of the eventual death of the universe itself, that is, the macrocosmic regression.

REFERENCES

Blacker, Carmen, *The Catalpa Bow: A Study of Shamanistic Practices in Japan*, London, 1975.

Clark, Charles Allen, *Religions of Old Korea*, Seoul, 1961.

Eliade, Mircea, *Shamanism: Archaic Techniques of Ecstasy*, translated by Willard R. Trask, New York: 1964.

Encyclopedia Britannica, Vol. XX, Chicago, 1973.

Hyŏn, Yong-jun, *Chejudo Sinhwa* (Mythology of Chieju Island), Seoul, 1976.

Illyŏn, *Samguk Yusa: Legends and History of the Three Kingdoms of Ancient Korea,* translated by Tae-hung Ha and Grafton K. Mintz, Seoul, 1972.

Lee, Du-hyun, *Han'guk Kamyon gúk* (Korean Masked Dance-Drama), Seoul, 1969.

Park, Pong-sok, *Kuksa Chonghae* (Concise History of Korea), Seoul, 1963.

TOMISLAV LONGINOVIĆ

1989: A Psychopolitical Analysis of the East European Revolution

It would be hard to deny the importance of the events that, sparked by Gorbachev's reforms, forced the military-bureaucratic machinery in the Eastern European countries to yield to the demands of oppositional political formations and start searching for compromise solutions. Without the support of the Soviet Union, the Warsaw Pact regimes proved to be a herd of paper tigers, a set of political structures alienated and detached from the people they have been oppressing for nearly five decades. The grip of the destructive and repressive political systems, supported by Red Army firepower, is now slowly weakening the centripetal force of the monolithic, one-party orbit for most of the previous 'satellites.' The Soviet leader's plan of radical restructuring worked much faster in East European countries than in the Soviet Union itself.

Mikhail Gorbachev's public relations effort has influenced public opinion to the degree that the majority of people in the West believes that 'communism is dead.' The misleading identification of the military-bureaucratic dictatorships in Eastern Europe with 'communism' is only one of the ways the media establishment interprets political events for the 'majority of individuals' in the West. But, while the Western media celebrates the triumph of capitalism and promotes images of an Eastern European amorous relationship with 'western-style democracy,' an old specter draped in new clothes is haunting the area. The abolition of one form of totality seems to be succeeded by another one, the totality of nationalism. The Marxist hope that the proletarian revolution will by itself solve the national question in Eastern Europe has proved wrong, since the failure of the oppressive political systems has demonstrated that under the thin surface of internationalist Marxist-Leninist ideology, the silent beast of nationalism had been quietly nurtured by the heralds of a multitude of small but proudly chauvinistic ethnic entities. The recent boom of literature which exploits the great national themes in Eastern Europe is related to a phenomenon that not only degrades the significant contribution of the twentieth-century giants from the area (Kafka, Schulz, Čapek, Gombrowicz, to mention only a few), but also creates conditions for the outbreak of potentially catastrophic conflicts among different nationalities in this already devastated part of Europe. It is within this context of shifting historical formations that I propose to examine the problems of Eastern European culture in general and literature in particular.

WHAT IS PSYCHOPOLITICS?

The psychopolitical analysis of literature insists on the irreducible connection between what apparently is an individual artistic expression in the text and the broader ideological and hegemonic context into which that creation is inevitably inserted and in which it is interpreted. Whether the author of the particular work of art (literature in this case) consciously or unconsciously embraces this or that ideology while creating the text is irrelevant. It is interplay between the psychological phantasms and the ideological frameworks within which the text realizes its meanings that above all informs the interpretation. This interplay is inevitably double-edged and undecided, since the agencies involved in the creation of meaning (the author, the text and the reader/interpreter) are equally subjected to the domination of the psychopolitical contexts within which the system of signs is transformed into a literary text. Instead of writing multiple obituaries for the writer, for the work and for the critic, this method affirms the subjectivity of each agency involved in the literary process by hollowing out the semiotic traces which persist in both the textual and the somatic configurations in the writing practice. It is obvious what this method owes both to materialism and psychoanalysis, with an added emphasis on the operations of language itself. But neither does it embrace the master texts of Marx, Freud, or Derrida: it does not attempt to propagate a principle that would replace the dominant role of the class struggle, the drive or the *differánce*. Conscious of its own inability to reach either the ultimately positive or the ultimately negative position in relationship to the text and its contexts, it therefore proposes one of the many possible readings of politics and its psyche.

AND NOW WHAT, MILAN KUNDERA?

One of the most productive examples for the theory of psychopolitics is the writing of Milan Kundera, whose role as the Czech national saviour (especially in the *Book of Laughter and Forgetting*, 1978) abruptly ended last year. Kundera's appeals to the enlightened West to save the small Bohemian state from the iron embrace of the Soviet monster-mother have a definite ring of anachronism after the events of 1989. The pseudometaphysical fabric of his literary text begins to collapse, now that the psychopolitical border is re-established between the Czechs and the Russians. At the same time, the rise of the Slovakian nationalism threatens to multiply the borders within Czechoslovakia itself, exacerbating the trend of small nations which demand to secede from the larger administrative units containing them. The increase in demands for independence demonstrates the degree in which nationalism can be appealed to when everything else, but primarily the economy, fails. Nationalism, like all other forms of paranoid discourse (racism, fascism) invokes a type of arch-ideology, something that is unquestioningly shared by a particular ethnic configuration. Now that the power of the bureaucratic-military dictatorships in the East has been replaced or reduced, either peacefully (e.g., Czechoslovakia/Havel) or by ritual slaughter of the old ruler (e.g., Romania/Ceausescu), we can conclude that the position of the writer will change significantly.

First of all, writers abroad are bound to lose their role of exiled oracles, since the context of political oppression in the East is apparently vanishing. Without that context, writers like Milan Kundera, whose entire poetics rests on the edifice of totalitarian oppression, are likely to find themselves in a literary vacuum. Works that have rightfully evoked indignation and sadness in the tender souls of Western liberals can now be interpreted within an entirely different context. Consider this lament of the Czech historian Milan Hubl, whom Kundera fictionalizes in *The Book of Laughter and Forgetting*:

> 'The first step in liquidating a people is to erase its memory', said Hubl. 'Destroy its books, its culture, its history. Then have somebody write new books, manufacture a new culture, invent a new history. Before long the nation will begin to forget what it is and what it was. The world around it will forget even faster.'[1]

The 'nation' in this case is a small, oppressed Czech nation which suffers the erasure of its national identity by the brutal and uncivilized Soviet Red Hordes, which invade Czechoslovakia from the East. Of course, the Slovaks are not even mentioned here, although they were not only dominated culturally by the

[1] Milan Kundera, *The Book of Laughter and Forgetting,* tr. Michael Henry Heim, (New York: Penguin, 1981), p. 159.

Russians, but by the Czechs as well. In view of last year's events, Kundera's lament gains an entirely different dimension. If before we could only sympathize and grow indignant over the predicament of Czechoslovakia, now we also notice that his statement is symptomatic of the new problem emerging in Eastern Europe. If the Russian nation is oppressing the Czech, then the only weapon the Czechs have at their disposal to fight back is also nationalism: 'its books, its culture, its history.' Although one could not call Kundera a nationalist writer, I have purposely chosen his case to show how deeply the problem of nationalism permeates Eastern Europe. Presumably, the only way it is possible to organize political and cultural opposition to the communist version of totalitarianism is the nationalist one. Even Kundera, the great spokesman for the European version of cosmopolitanism, is not entirely immune from this ailment. Draped in the cloak of the French intellectual fashion of the day, this master of literary technique, of Cartesian doubt and postmodernist irony, reveals himself as a closet Czech nationalist.

Milan Hubl is mourning the 'liquidation of people' which in turn results in the erasure of one's language, one's literature, and one's own personal identity. Identity is tacitly established by Kundera as something dependent on a nation, and in his own work that (Czech) nation is constantly being victimized. The relationship of the pseudofeminine 'Mother Russia' with her politically dependent infants is the main focus of Kundera's last three novels (*Life is Elsewhere, The Book of Laughter and Forgetting* and *The Unbearable Lightness of Being*). The first one, which deals with what Kundera calls 'the lyric attitude,' is a parodic *Künstlerroman* about Jaromil, a mummy's darling who wishes to become a poet. He becomes one, through collaboration with the police and the party authorities. Throughout the novel, the power of Jaromil's mother (who is never explicitly named) is equated with that of the repressive political system, supported by the Soviet Union. In *The Book of Laughter and Forgetting*, Marketa, an ugly and possessive woman, is identified with Husak's regime of forgetfulness. And in his latest book, Kundera juxtaposes the characters of Teresa, who is possessive and jealous of the main protagonist Tomaž, and Sabina, who is cynically free from such attachments. Of course, Teresa is the one who cannot bear exile (with its unbearable lightness of being) and returns to be oppressed in Czechoslovakia. Most of the female characters (with the notable exception of Tamina from the *The Book of Laughter and Forgetting*) are somehow identified as weak and incapable of taking a firm moral stand against the totalitarian regime. They are implicated as the ideal collaborators of the communists, since they lack a sense of identity distinct from that of the military-bureaucratic power.

The Russians have replaced direct physical confrontation by a public relations campaign, an ostensibly Western mode of political control, which disregards the binary opposition between East and West and seeks to mesmerize the public in ways which are uncannily similar to those of the writers. (Is that why writers are turning out to be the most successful political candidates in the East?) The question remains open: how is Kundera's new novel, entitled *Immortality,* going to respond to the changes in Eastern Europe? Since his poetics of the novel so heavily depends on the repressive nature of the Soviet regime, how is he going to respond to the change in tactics introduced by Gorbachev?

BLOOD IN THE BALKANS?

Although Yugoslavia was not directly influenced by Gorbachev, events in the rest of the communist world echoed quickly in this region famous for its instability. While Romania and Bulgaria have experienced a degree of cosmetic reform, Yugoslavia seems to be on the verge of collapse, due to a similar, basically nationalist threat. The cult of the Great Father, Marshall Tito, is being attacked from all sides, with a violence and aggression that only shows how great political repression under his regime really was. The press is exposing his lavish lifestyle reminiscent of Latin American *caudillos* (including the quantity and type of his underwear), while historians are racing to document how he has harmed this or that nationality within Yugoslavia. It seems that the entire country has suddenly begun to deal with a collective Oedipus complex, which has resulted in a rejection of the Great Father and identification with a particular national ideal.

Writers find themselves in a situation which favours those who have made a clear national choice. Those who feel that nationalism represents a dangerous flirtation with a long history of hostility, especially between the Serbs and the Croats, are running the risk either of being labeled defenders of the Titoist version of autocracy or defeatists who are naive enough to believe that a new concept of Yugoslavia is possible after communism. Both nationalist leaders, Franjo Tudjman in Croatia and Vuk Drašković in Serbia, are writers, or at least belong to the respective Writers' Unions in Zagreb and Belgrade. The former has already won the presidential election in Croatia, on the overtly nationalist platform of his HDZ (Croatian Democratic Union). The latter, Vuk Drašković, has taken on the Bolshevik strongman Slobodan Milošević and is expected to win the first free election in Serbia in December of 1990. His SPO (Serbian Restoration Movement) is trying to wrest the nationalist initiative from Milošević, who won popular support among the Serbs with his military-police actions in the unstable province of Kosovo.

A situation in which the balance of political powers shifts daily, and where the Great Rock-and-Roll Party (which wants to nominate a female bass guitarist for President) can run for office alongside the New Royalist Block (which wants to reinstate the monarchy and bring the Prince of Yugoslavia from London), seems so absurdly surreal that a writer who relies on the imagination and invention cannot but feel a bit superfluous. With the readership visibly diminishing, due to the astronomic prices of books and the aggressive and sensationalist journalism, writers whose concerns are primarily literary are once again in the position of an endangered species. Pavao Pavličić, the best-selling Croatian author writes:

> The most disappointing thing in this new situation is exactly that: I understand the importance of the moment and the fact that the worlds are collapsing and the new worlds are created; a bit of history here, a bit of history there, but I have always thought that in our profession the majority of people consider literature the most important thing in their lives. It turns out that I was wrong. Luckily, it's not the first time this is happening, so we are used to it.[2]

If literature is once again transformed into the handmaiden of politics, we may be witnessing the rebirth of soc-realism, which this time could bear a more appropriate label of nac-realism. Nobody has yet come forward with guidelines for writers similar to those of A. Zhdanov and it is very unlikely that this will be necessary in a universe permeated with the demands of nationalist ideology. The censorship here is almost automatic, since nationalism is perceived as part of one's personal identity, not something enforced from the outside like revolutionary socialism. But, censorship exists, as the case of the satirical supplement for *Nedeljina Dalmacija* which was recently banned by the Tudjman government in Croatia shows. The promises of tolerance and pluralism in a new, democratic Croatian state were jeopardized by this politically unsound move.

Across the border, in Serbia, a literature that glorifies the medieval Serbian kingdom, or laments Serbia's dead in two world wars, is becoming absolutely dominant. (Ćosić, Smiljanić, Bećković are only the most notorious in this trend). This is what those who can still afford to buy books in Serbia crave: restored

[2] Letter received from Pavao Pavličić, May 1989.

national pride in times of economic hardship and political uncertainty. Anti-Croatian writing in the press is also on the rise; an Orthodox priest who writes a column for the SPO newspaper (*Srpska Reč*) recently published a vicious etymological piece on the word *Hrvati* (Croatians). The conclusion: the Croatians are a people whose name reveals their true, bloodthirsty nature. This lay analysis relates the sound of the word blood (*kry*) to the first syllable of that name (*hrv*).

The effects of this type of reasoning are abundant on both sides of the Serbo-Croatian border, a border that divided the Roman Empire into its Eastern and Western halves and continues to divide Yugoslavia today. Croatian nationalists feel that they are all that stands between the West and the dark, Byzantine non-culture which they perceive in Serbia, Montenegro and Macedonia. The Catholic religion and the Latin alphabet are the external signs of their alliance with the West, while Christian Orthodoxy and the Cyrillic alphabets are seen as the signs of cultural and historical otherness.

On the other hand, Serbs have a culturally built-in messianism, which convinces them that they are the leading nation in the Balkans. This messianism stems from the imaginary connection with the medieval kingdom of Tzar Dušan, which once extended from Transylvania in the north to the Corinthian straits in the south. The destruction of the Serbian empire by the Turks, in 1389, at the Battle of Kosovo (the same place where the Albanians are now rebelling) is a historical marker connecting Serbia with the West. This loss is perceived in the popular imagination as the sacrifice of Serbia, a Christian nation, for the rest of Europe, which is thus saved from the invasion of the Ottoman infidels.[3]

[3] It is interesting that a similar rhetoric of sacrifice was used after the Second World War, to prove that Yugoslavian partisans started the revolution in 1941 in order to prevent the Germans from reaching the Soviet Union before the winter. This is just one of many examples of the way communism has

In 1989, the secentenary of the battle was used by the communist leaders, headed by Milošević, as a means of assembling more than a million Serbs at the famous spot, in a powerful propaganda gesture. Most of the Serbian national identity is related to wars and the military. A popular song illustrates both the messianism and the militaristic quality of Serbian national identity:

> Ko to kaže, ko to laže
> Srbija je mala
> Nije mala, nije mala
> Dvaput ratovala
> I opet će, i opet će
> ako Bog da sreće . . .

The translation conveys only a part of the emotional charge of this song: 'who is saying, who is lying/that Serbia is small/It's not small at all/It fought two world wars/And it will fight again/If God grants some luck . . .'

This sort of attitude tends to transform non-Serbs into wimps who have never had enough strength and courage to fight and liberate themselves as the Serbs did. In fact, Serbian messianism is supported by their historical role in the formation of Yugoslavia, which is perceived by the other nationalities as a Serbian hegemony and unjust domination. So, the popular Slovene slur for Serbs is 'Turks' or 'Gypsies,' a qualification that reveals how much the 'European' identity of one Yugoslav nation depends on its ability to deny its own obscure Balkan roots and charge the other (Serbian, Macedonian) nation with homelessness (Gypsies) or with Asian origins (Turks). This is only a part of what constitutes the contemporary political madhouse of Yugoslavia, a confusion which could easily end up in civil war. But some, including the Serbian/Jewish author David Alabahari, are more optimistic:

---

appropriated Christian rhetoric.

It is really hard to explain what is going on here. However, one thing is clear: the communists are panicking, especially here, in Serbia, where they are still formaly in power and where they have not yet tasted the bitterness of defeat. After the communist boot, nationalism is a natural thing. The same thing has happened in Bulgaria, Romania, Poland, Russia. The easiest thing is to be a nationalist. The most difficult thing is to have a brain, since it bends the spine and makes the head feel heavy. In my opinion, a bit of patience is necessary. The passions will calm down—which is already the case in Slovenia and Croatia, after the nationalist euphoria caused by the first free elections—and brighter people will remove the madmen from the throne. Yet, the only thing that could unite us is some supranational idea, something that is not very popular these days.[4]

The slightly ironic tone of the letter shows a certain detachment which Albahari, as a person whose primary interest is literature, has to maintain if he is to continue functioning in the highly politicized universe of Eastern Europe. His own fiction, apparently disengaged from overtly political solutions, is a perpetual exploration of the process of writing and its relationship to personal identity and experience. It may strike one as unusual, but the fact is that both Albahari and Pavličić (who writes market-oriented prose) belong to the newly formed Independent Writers' Union. Although their styles and literary interests, as well as national and cultural alliances, are sharply different, they have joined the pro-Yugoslavian Union in order to show that there are still people who believe that writing can survive as an autonomous activity in post-communist Yugoslavia.

IS THERE A FUTURE?

Philip Roth, the tireless American promoter of East European writing, once said that it was because nothing was allowed there that everything mattered. In the West, he lamented, everything was allowed and therefore nothing really mattered. In this case,

[4] Letter received from David Albahari, 20 June 1990.

greater freedom in the East is likely to bring a diminished significance for writers and intellectuals, who are traditionally the ones to expand the limits on poetic freedom imposed by the Party control apparatus. This may be true only to a certain extent: political events are evolving with such a speed that readers seem to prefer journalistic writing to texts of artistic value. Since the ability of literature to provoke politicians is now almost entirely a thing of the past, writers are either entering politics themselves (e.g., Vaclav Havel) or trying to compete with the press in producing sensationalist literary texts, usually along the already-mentioned nationalist lines. We can also expect a boom of commercialism in literature, which is only a logical consequence of the introduction of the much praised concept of the market economy. We arrive at one of the many East European paradoxes here: while oppressive political systems stimulate morally sound, quality resistance, permissive ones produce insignificant and disposable literature for consumption.

I personally hope that there will always be those who will be able to grasp and criticize the new totality (nationalism) after the last traces of the communist bureaucracy are gone. Women, ethnically marginal groups, and those still in exile are the most likely candidates, since their vision occupies the marginal position necessary to counter the views of the compact nationalist majority, which seems to be gaining the upper hand in the region. The threat of rising anti-Semitism, the moves of the Catholic Church to ban abortion in Poland, Croatia and other countries that it dominates are real enough for those who are able to articulate their responses to oppression. Those in exile, on the other hand, have a clearer vision because of their detachment; it is likely that they will not be able to accept uncritically the substitution of one form of totality for another. If the writers in Eastern Europe are able to survive the economical and the ecological catastrophies of the region, we may soon be catching the literary echoes of the 1989 revolution in their work.

JASON WILSON

Literal Translations: Recent Spanish American and Spanish Poetry in English Translation.

What poetry crosses over into English from the Spanish-speaking world tends to follow some political or cultural event that breaks into the media; like winning a Nobel prize, or a revolution, or the horrors of repression, or the impact of feminism. Outside such events an English-speaking reader will have trouble savouring the diversity of poetic talent given the amount of small presses, the poor distribution, and isolation of Latin American Republics from each other. However, if one can generalize about the late 1980s, an alert reader can still detect two literary currents. First, there is still a sense of an avantgarde; what Octavio Paz has called a 'tradition of rupture'; and second, thanks to history, there is a prevalent literal poetics of political outrage, joined now by the feminist rediscovery of women poets.

Ariel Dorfman, born in Argentina in 1942 but a Chilean national, is a case in point. *Last Waltz in Santiago and other Poems of Exile and Disappearance* (Penguin: New York, 1988) is translated into English by Dorfman with Edith Grossman, and is not bilingual. With direct literal poetry of this kind a bilingual edition is less necessary, for language is secondary to the purpose of awakening, reminding, and changing the reader. There are few explicit references to the poetic tradition, while the title ironizes Bertolucci's film *Last Tango in Paris.* It is aggressively Latin American—tangos come from Buenos Aires and waltzs from Vienna—and solves the problem of demonstrating that Dorfman is a Latin American poet. 'Poems of Exile and Disappearance'; no hiding behind metaphor or conceit; these realities are named directly and simply.

Dorfman faces the problem of saturation. The news media have over-exposed the evils of Pinochet; articles, documentaries, books, testimonies and reports have remorselessly piled up. At the same time the events of 1973 in Chile are fading into history; a whole generation of readers has been born since Allende's murder. In an author's note Dorfman picks on two expressions that 'stubbornly resisted an English-language equivalent': one is Brazilian, *pau d'arara* and the other, *compañero.* The latter 'could be rendered as mate, friend, comrade, companion but none of these has the unique resonance of the Spanish'. This key word enshrines a uniqueness that Dorfman does not want to lose in translation. Left in Spanish it retains its emotional force.

Dorfman calls his position in exile one of 'interpreter'; letting others speak through his words and resisting the way lyric poetry spins off into magic and incantation: 'please forgive any rhymes or rhythms you may find'. A poem ends: 'Let them speak for themselves'. He spreads his poems out over the white page to emphasize phrases, and repeats passages to underline the emotions. Even political poetry aiming at immediacy cannot avoid rhetoric. The poems ask a reader to 'think about it' and 'understand', discarding official explanations, reading between the lines. From exile, Dorfman recreates long-distant telephone conversations, and weeps over soap operas on TV, but not over the news of a friend's death. The best poems in this collection deal with painful emotional paradoxes, the differences between conscious and emotional responses: 'I know the tricks and techniques of the movies/ I've studied how the violins manipulate us/ I spent my life denouncing Doris Day / but during General Hospital / something clouds over inside / and something wet and salty runs down my cheeks.' The best poem in the collection 'Sun Stone' manages in its direct language to be both moving and painful; a prisoner and a soldier—both from the pueblo—discover they are on the same side '–Forgive me, compañero–' but have to do their duties. The 'echo' of the soldiers' commiseration helps the prisoner 'not hear / the sound of the shots'.

Dorfman exploits a Latin American tradition based on sincerity, that starts with César Vallejo and branches into Gelman, Cardenal, Roque Dalton and many others. Dorfman writes: 'Let's tell the truth once and for all', or 'I write down the facts carefully', or deals with moral problems: 'But how can I tell their story / if I was not there?'. He never exposes his own poetic cleverness, and necessarily remains literal, and easy to translate. He also avoids slang or jargon (unlike say, Dalton). Few expressions jarred on my ears (the only one was 'perk up your ears'). The problems that face a translator are cultural/historical ones; how to explain the unpleasant associations of Villa Grimaldi, or the 4th of September 1973. Much of the emotion of these poems depends on a prior knowledge, and even political empathy.

Ernesto Cardenal became Minister for Culture when the Sandinistas came to power after Somoza was driven into exile in 1979, and later assassinated in Paraguay. As Cardenal moved from Trappist or Merton leanings to guerrilla action, so did his poetry become increasingly politicized. Thanks to the immediacy of what poetry voiced there could be no delving into the subconscious, no appeal to dark metaphors. The poem had to be literal. Cardenal's access to his readers was reading his poems aloud, bypassing the dependence on books when both paper and literacy were at a premium. But his poems have ended up in books, without the appeal to his dramatic way of reciting. A character in Mario Vargas Llosa's novel *The Real Life of Alejandro Mayta* (trans. Alfred MacAdam, Faber, 1986) says this about

Cardenal: 'he made a vivid impression on me, too. I haven't said that I'll always be sorry I heard him, because since then I haven't been able to read his poetry, which I liked before . . . Everytime I try to read him, something like acid flows out of the books, and ruins it—the memory of the man who wrote it . . '. He lists various reasons—insincerity, hypocrisy, demagoguery, dressing like Che Guevara. Whether you agree or not with this view, Cardenal's poems do change when read aloud. How can a translator convey this extra-literary function in a translation?

Dinah Livingstone follows her *Nicaraguan New Time* (Journeyman: London, 1988) with a secuon from a long epic Cardenal has been writing during his years in political office, called *The Music of the Spheres* (Katabasis: London, 1990). She sticks close to the literal vocabulary and syntax, refusing to embellish, or create another kind of poetry in English. Clearly she wants Cardenal's didactic plainness to cross cultures and language, not create a separate poem in her English. The bilingual edition allows a reader with a minimum of Spanish to leap back and forward between the texts. Only in a few places did I sense that Dinah Livingstone tried to improve Cardenal's slightly monotonous quality. Cardenal links his co-operative, dialectical, political views with science: 'When this was corrected, the universe stopped being static / and began to tick'. The rhymes spring from the English, not the Spanish line endings: ' . . . estático / . . . . movimiento'. At another point she adds verbs for cocks to 'crow' and frogs to 'croak' when in Spanish they simply wake and go to sleep. But generally her faithfulness does not bring this didactic poem to real life. It awaits a voice, being recited.

Whilst living in New York in the 1940s Cardenal discovered, and translated, North American poetry, especially Ezra Pound's *Cantos*, and the way history was incorporated into lyric poems. Cardenal's strange attraction within the Hispanic tradition has its roots in his borrowings from Pound; his poetry of the 1960s achieved a marvellous diversity of tones. When the Peruvian poet Antonio Cisneros (b. 1942) won the Cuban Casa de las Américas poetry prize in 1968, a new kind of poetry came to the attention of readers excited by Che, Mao and leftist liberation movements, Cisneros followed Cardenal by not adapting French surrealists, or rehashing Pablo Neruda; he sounded different, wrote a narrative, mocking poetry, with sudden shifts, juxtapositions, college quotations and colloquialisms, reminiscent of Pound and Ginsberg (he later translated 1960s British poetry). *At Night the Cats* (Red Dust: New York, 1985) is a bilingual anthology of his work from 1964 to 1981.

Cisneros's poems display a peculiar sense of being a stranger in his own country, at odds with the official histories, and uncomfortable with the rich Inca past. He illustrates Pound's 'botched civilization' from a Peruvian point of view. This adoption of North American models makes it easier for his translators, yet he sounds tamer in English. Cisneros lived for a while in Southampton, England, refusing to be seduced by European glamour. Life here left Cisneros shut 'inside my box of Cornflakes / writing for the sheer hell of it'. The three translators—Maureen Ahern, William Rowe and David Tipton—have persevered with Cisneros through his many styles over the years and through several books. Their translations stick close to the free-flowing, rhymical Spanish, as this is part of an International Style. In most cases where my eye wandered from poem to translation I noticed a careful literal English, as if Cisneros' poems were already partly English. This generous selection ends with David Tipton's note 'Translating Cisneros', but he does not analyse the problem a translator is faced with when a poet sounds 'unusual' in Spanish because he has read idiosyncratically into an English poetic tradition of un-Spanish irony and precision. His note is more about the history of translating Cisneros over fourteen years of collaborative efforts.

The Argentine poet Roberto Juarroz (b. 1925) writes quite a different kind of poetry. His tradition is French, a mental poetics that questions language itself in endless riddles and games, close to Heidegger's philosophy. Juarroz does not load his words with local colour, nor proclaim his South Americanness; he is Platonic, reaching beyond the arbitrariness of language as a system, so that you sense an Ur-language at the edge of the actual poem, whatever your mother tongue. Translating Juarroz into English is extremely difficult as English is allergic to abstractions, craves precision, particularities. Juarroz is a great reader of his quiet poems; his booming voice comes as a surprise for you expect him to whisper, but it does give body to his short poems that sometimes does not come from the page. Yet Juarroz has been well translated in French. *Vertical Poetry* (North Point Press: Berkeley, 1988) collects W.S. Merwin's translations from Juarroz's nine books of 'Vertical Poetry'—like the Spaniards Luis Cernuda, and Jorge Guillén, each new book retains the first title. Juarroz also sounds different in Spanish: apolitical, un-surrealist, dealing with mental and perceptual paradoxes, verging on religious poetry.

Again W.S. Merwin can only be faithful, and literal, despite the security of the Spanish in this bilingual edition. Juarroz does use strong verbs which Merwin matches; 'Sometimes death grazes our hair, / rumples it / and does not enter. / Some great thought perhaps giving it pause? / Or can it be that we are thinking / something greater than thought itself?' The actual words are so plain because they set off thinking about thinking. But note 'grazes' for 'roza' and 'rumples' for 'despeina'. This poem depends on the verbs. My favourite poem comes out well in Merwin's English: 'In the gut of summer / like a clear thread / echoes the voice of the ice cream seller. / It is not childhood coming back. / It is not something divine dressed in white. / It is not the moon in daytime. / It is only the possible / showing us that it exists. / The impossible never raises its voice'. The main reason Juarroz translates well into English is that he sounds dissonant and odd in any language.

Another angrier literal discourse can be discerned in the late Rosario Castellanos' (1925–74) poetry. She has been rediscovered, even in her own Mexican culture, thanks to feminism's redressing of literary history. Castellanos ranged through various genres—novels, journalism, plays etc.—but began as, and saw herself essentially as, a poet. She linked the suffering of Indians in her home state of Chiapas, with the suffering of women, and was able to relate her work to a greater whole than simply writing good poems. The bitterness of both the Indian and woman's lot sharpens her need to communicate through direct language. This should make it easy for her translators. The pitfalls are turning her into a 'mystic' and sentimentalizing her insights.

Myralyn F. Allgood's *Another Way to Be. Selected Works of Rosario Castellanos* (The University of Georgia Press: Athens & London, 1990) comes as an academic work, with full bibliographies and erudite notes, a work of love and respect. It aims at representativity, with passages from her fiction and journalism. The poetry selection is short, but bilingual. When the poem strives for emotion I do not like the English translation. For example, the prose poem 'Indian Mother' ends 'Oh, may sleep elude our brows, may friendship pierce our hearts like thorns, and may songs offend our mouths . . . ' when the Spanish is far stronger; 'desampare' is to abandon not elude, 'ortigas' are stinging nettle not 'thorns', 'pudrir' means to rot, not 'offend' our mouths etc. I contrasted Myralyn Allgood's version of Castellanos' angry poem about the student massacre at Tlatelolco with Maureen Ahern's (first published in *Looking at the Mona Lisa*, 1981, and then reworked in *A Rosario Castellanos Reader*, University of Texas Press, 1988) and from the opening tercet you sense that Ahern is a poet:

Darkness engenders violence
and violence seeks darkness
to cover its crime (Allgood)

Darkness engenders violence
and violence invokes darkness
to jell in crime (Ahern)

Note the weak 'seeks' and 'cover' for 'pide' and 'cuaja'. Even in literal translations there is work to be done to keep the tone right, not let the translation sink into poeticisms.

Maureen Ahern and Magda Bogin's *The Selected Poems of Rosario Castellanos* (Graywolf Press: Saint Paul, Minn, 1988) overlap with the crucial poem 'Meditation on the Brink' and both versions remain powerful. Contrast these openings:

No, it's not a solution

to throw oneself under a train like Tolstoy's Anna
or gulp down Madame Bovary's arsenic

or await on the barren heights of Avila the visit

of an angel with a fiery dart
before binding one's veil back over one's head

and starting to act (Ahern)

No, throwing yourself under a
train like Tolstoy's Anna

nor hastening Madame Bovary's
arsenic
nor waiting for the angel with
the javelin
to reach the parapets of Avila

before you tie the kerchief to
your head
and begin to act (Bogin)

While not invoking what Alastair Reid once called the 'Translation Police' Bogin gets 'páramo' wrong (it's not a parapet); did Madame Bovary 'hasten' down arsenic or 'gulp' it down? Nancy Mandlove translated the same poem in *The Renewal of the Vision: Voices of Latin American Women Poets 1940-1980* (ed. by Marjorie Agosin and Cola Franzen, Spectacular Disease, 1987) where Emma Bovary 'downs' arsenic. Ahern knows that Avila is high and barren, not 'the barrens' (Nancy Mandlove), and that people wear veils there not cloaks, or kerchiefs. But the comparisons are odious, and both translators bring a real Rosario Castellanos into English. I am staggered, and guilty, that Castellanos is only now being brought to English-reader's attention.

Both Ahern and Bogin explore the problems of translating Rosario Castellanos. Ahern raises the crucial point: 'The quintessential Mexicanness of her discourse challenges the translator. Because so much of the author's work is culture bound, cultural equivalences take precedence over lexical ones'. Ahern narrows this general point down to the following assumption: 'a great deal of cultural information that is automatically shared by the Spanish-language readers' (doesn't she mean Mexican-language readers?). This prompts her to advocate notes as an 'integral part' of her translated texts, a point I share. Ahern also mentions the 'intense orality' of Castellanos's work, with its inversions of middle-class Mexican clichés and idioms.

Bogin's remarks centre on the translation of tone. She has sought to retain from Castellanos's Spanish 'what I can call the gently elevated, at times even heroic, feel of Castellanos' diction' and leans heavily on 'romance' words like 'realm' instead of 'kingdom', hoping to resolve the problem of Castellanos's 'flat' style, because a 'flat' style in English is different to a 'flat' style in Spanish. However if Bogin errs it is into interpretation. In a wonderful poem about the pre-Columbian dead where Castellanos can only gaze at trees above ruins and betray the past, we read in Spanish about a wind 'passing' that 'bites' with acid teeth. In Bogin this becomes too lyrical: 'acid teeth *gnash* / the wind as it *streams* by'. When Bogin does not wax lyrical she conveys Castellanos well, though I preferred Ahern's more vivid English, epitomised in her rendering of the title 'Meditation on the Brink' for 'Meditation on the Threshold' (Bogin).

If we move from literal poetics—political, feminist—to experimental and metaphorical writing a different set of problems occurs. Viking Penguin has just published the third translation of Federico Garcia Lorca's *Poet in New York* (1989), with interesting notes and appendixes, aimed at students. Comparing Greg Simon's and Steven White's new versions with earlier ones by Rolfe Humphries (1940) and Ben Belitt (1955) there is no doubt about the validity of another attempt at matching Lorca. I feel that the earlier translations have dated while we are closer to Lorca's nightmare vision of the City. However, the two admirable translators have no say in this edition: I would have liked to have known how they worked. Did one prepare a literal version bettered by the other etc? Did they work from the earlier editions? It is a pity that this book is aimed at readers who want a maximum of information, and not at readers of poetry.

From the opening poem 'After a Walk' you see what the translators have achieved with Lorca's Spanish by exaggerating his words. They have created ominous poems in a bilingual edition not through rhythm, repetition and accumulations, but lexically. This poem opens in Spanish with 'Asesinado por el cielo', turned into English as 'Cut down by the sky' where the sense of being dwarfed—and killed—by skyscrapers echoes Lorca's first impressions. Lorca's 'formas que buscan el cristal' (literally 'forms that seek glass') becomes 'crystal-craving shapes'; the white-egg face of a child becomes 'the blank face of an egg'; a 'broken head' becomes 'whose skulls are cracked'; 'with everything that has a deaf-mute tiredness' becomes 'with all the bone-tired, deaf-and-dumb things'. Nearly every poem in this edition testifies to this heightening of tension. Cats do not 'eat' but 'devour' frogs; 'bits' of lemon are 'rinds'; staircases, 'fire-escapes', hitting, 'swatting' ... With the Spanish facing, you can move into this white-hot English where general categories in Spanish like 'escalera' (i.e. *any* staircase) become particulars, 'fire-escape', and more New Yorkan.

You feel that the translators have got inside Lorca, and recreated their sense of him in English. Another crucial poet from the 1920s and 30s is the experimental Chilean Vicente Huidobro (1893–1948) who, when he arrived in Cubist, First World War Paris, changed his language, and wrote poems in French. His long seven canto poem 'Altazor' was begun in French in 1919, and finished in Spanish in 1931. Eliot Weinberger has translated the complete poem in a bilingual edition as *Altazor or A Voyage in a Parachute* (Graywolf Press: Saint Paul, Minn., 1988). Huidobro invented his own school of poetry, based on images, and called 'Creationism'. This gave birth to a theory which decided that translating images was easier than translating words. But by the 1930s Huidobro's poetry began in a quasi-surrealist way 'frolicking with magnetic words'. By the fifth canto Huidobro is breaking words up, and trying to create a new language from the segments and fragments. Sometimes Weinberger matches this 'thrill' as in the play with mandolin and farandolin (a French dance): 'La faranmandó mandó liná' becomes 'The faramandole that manned a linn', but by the seventh canto Weinberger is unable to match Huidobro. For example, the line 'Hormajuama marijauda' becomes 'Mutopiform matropia'. Why? The Spanish plays with 'hora', 'marijuana', 'mar'(sea), 'damajuana' (demijohn) and 'Juana' but I cannot see this in the English. Huidobro here is untranslatable because he intends (and fails) to create a 'new' and 'universal' language, but it only plays with cosmic associations. However, Weinberger has made a brave attempt that stretches from his good, literal renderings to free and faithful combinations at the end.

Weinberger explains his decisions to avoid notes, and stick literally to the Spanish, and adds: 'although this is a translation and not an "imitation", it is quite possible that another English version would be utterly dissimilar to this one . . . The game of *Altazor* has an infinite number of moves, and I look forward to its next translation match'. But he really means the last three cantos, for the first four are there in his literal version, and would not change if translated by another. Avant-garde poetry also demands a literal version, unless it breaks into punning. Huidobro formed part of a branch of Hispanic writers, with Olivero Girondo and César Vallejo as his word-playing equals. The former has not been translated.

However, a new translation of the seminal Peruvian poet César Vallejo, *César Vallejo: A Selection of his Poetry* (Francis Cairns, 1987) raises the question of identification between poet and translator by reminding me of an earlier and ambitious translator of Vallejo, Clayton Eshelman. In the introduction to his translation of *The Complete Posthumous Poetry* (University of California Press, 1978) Eshelman, with José Rubia Barcia, describes his agony with the dead poet, its 'psychic as well as linguistic ramifications' where 'Vallejo got into every corner of my Being so thoroughly as if 'the man I was struggling with not only did not want his words changed from one language to another but seemed as if he did not want to be changed himself'. The latest translator of Vallejo does not follow this agonised trial of translation. James Higgins has modestly compiled an anthology with plain translations and interesting notes in a bilingual edition where his English is merely a path into Vallejo, not something that can stand on its own. Sometimes his faithfulness betrays Vallejo as in the opening of the radical poem XXXIII from *Trilce* where we read 'Warm oven of these biscuits of mine / pure countless essence of childhood, mother' where the Spanish is 'yema', egg yolk, not the weak 'essence'. What I sense behind Higgins' approach is that translating Vallejo is impossible, but that a literal version is better than Eshelman's ego-centred agonistic version. After all, Vallejo himself did not believe in translation of poetry, poetry being untransferable, a 'cardiacal tone'.

When academics translate they produce long introductions, footnotes, and bibliographies so that the work can figure in a list of their own publications, a format evident in several of the books under review. Jack Schmidt's translation of Raúl Zurrita's second book in English, *Anteparadise* (University of California Press, 1987) follows this format. Zurrita is an experimental poet who plays with an easily translatable area, that of playful topography, with lists. He wants to break the mould of the old lyrics, and includes photographs of his poems written in the sky. The topic of his long, linked poem is an emotional patriotism, a love for a Chile suffocated by Pinochet's dictatorship, with sections on Chile's natural beauties, its deserts, mountain ranges, beaches and people, akin to Neruda's more lyrical celebrations of the same elements. Zurrita's literary model sounds strange to Spanish ears as it is Biblical, the Song of Songs (and perhaps John Clare), verging on prose, with thematic repetitions. The effect is not lyrical but derives from his piling up of qualities. Schmidt has included notes and explains why he chose not to translate the Spanish 'Ay', or 'Cielito' or 'vidita'. At times Schmidt's attempt to catch the Biblical flow sounds awkward with 'Awaken' and 'Go forth' and 'behold' when the Spanish is 'Wake up', 'leave' and 'see'. He sticks close to Zurrita's inverted syntax—'And weeping she said' or 'but arid these landscapes were nothing but the evanescent Chilean landscapes'—when Zurrita's parodic diction comes from a language that easily inverts while English doesn't, or if it does, sounds archaic. In another note Schmidt explains how he solved the problem of Zurrita's play with gender—'un madre' for 'una madre' with 'muther' (for mother)—adding that he was 'striving for the same effect', but surely not achieving it, for how does gender change by substituting 'u' for 'o'?. Still, it is good to see such a different voice emerge from Chile's rich poetic tradition.

Finally, two books that have appeared over the last years collect together several earlier editions. The Mexican poet José Emilio Pacheco's *Selected Poems* (New Directions, 1987) has been edited with a mimimal of notes by George McWhirter, and includes versions by Ed Dorn, Alastair Reid, and others. In his short foreword McWhirter records that he sought more than a word for word translation, trying to capture Pacheco's sense of form, sometimes with the help of Pacheco—an excellent translator of T.S. Eliot and others from the English. What we have now is the chance to revalue those Mexican poets writing in the shadow of Ocatvio Paz. Pacheco is particularly apt, for he follows Paz's cosmopolitan lead, but writes lyrically from a demythifying, and ironic stance. Another poet who has dropped out of the limelight, and who works within the same dissident, and Anglo-Americanish tradition, is the Chilean Nicanor Parra. David Unger has edited *Antipoems: New and Selected* (New Directions, 1985), bringing together the best of Parra's work as an antidote to Neruda and the populist lyrics that derive from his position. Anybody wishing to redefine the varied Latin American poetic tradition could well start with Nicanor Parra, and then move into Pacheco and Castellanos, rather than the usual Neruda, Vallejo and Paz. In cultures where poetry is still read, and where personal and social issues still generate public passions, the poet has a place.

Notes on Authors and Translators

MARVIN BELL teaches in the University of Iowa's Writers Workshop. The author of eight collections of poetry, his most recent collection, *Iris of Creation,* was published in 1990. He regularly publishes essays in *The American Poetry Review,* among other journals. In 1983 he published *Old Snow Just Melting: Essays and Interviews.*

MIRON BIALOSZEWSKI was born in Warsaw in 1922, where he remained during the Nazi occupation, being deported by the Germans after the defeat of the 1944 Warsaw uprising. After 1955 he became well known for his avant-garde theatre as well as his poetry. He was regarded by many as the most innovative of post-war Polish poets. His *Memoir of the Warsaw Uprising* was published in 1970. He died in 1983.

KATARZYNA BORUN-JAGODZINSKA was born in Warsaw in 1956. She has published four collections of poetry. She lives in Warsaw with her husband Andrej Jagodzinski, a translator of independent Czech literature, and their two children. She is currently at work on a novel about life in Poland during martial law. In 1989, she participated in the International Writing Program at the University of Iowa.

JOSEPH BRODSKY won the Nobel Prize for Literature in 1987.

RICHARD CHETWYND is a graduate of the University of Iowa Poetry Workshop. His most recent chapbook, *Dozen Poems,* was published by Jezuum Cow Press. In 1988, he received the East European Poetry Translation Prize from Mr Cogito Press. He and

Grzegorz Musial have collaborated on a translation of a collection of Musial's poetry.

MANGALESH DABRAL was born in Tehri Garhwal, Uttar Pradesh, in 1948. A journalist by profession, he has worked for several newspapers and magazines, and is at present with *Jansatta* (Delhi). He has published two collections of poetry. In 1991, he participated in the International Writing Program at the University of Iowa.

TOSHIKO FUJIOKA was born in Japan, in 1947. She studied Japanese literature at Konan University and is a member of Japan's Poets Club. She is a peace member of the special branch of the Co-operative Society in Osaka, where she lives. She is one of the editors of the monthly magazine *Gekkan Kinban* and is also a 'coterie' member of the quarterly magazine *riviere.* Toshiko Fujioka has published three collections of poetry: *Morning after Rain, View,* and *City/Fish/Scenery.*

JAMES S. HOLMES, see headnote by Daniel Weissbort.

KAZUA HONDA was born in 1933 and graduated from Waseda University, majoring in English literature. He works in Irish literature, in particular on James Joyce. He teaches at Bunkyo University in Saitama, Japan.

ATTILA JÓZSEF, see introduction by Lucas Myers.

THERESA KI-JA KIM is Associate Professor of Theatre Arts at the State University of New York, Stony Brook. She is a specialist in comparative theatre of the East and West, and the shamanic origins of theatre.

ARUN KOLATKAR was born in Kolhapur, Maharashtra, in 1932, and educated at the J.J. School of Art, Bombay. He writes in both English and Marathi, and although his poems have appeared in magazines and anthologies since 1955, he has so far published only two books, *Jejuri* (1976), in English, and *Arun Kolatkarchya Kavita* (1976), in Marathi. In 1977, Kolatkar won the Commonwealth Poetry Prize. He lives in Bombay.

LALLA, see headnote by A.K. Ramanujan.

TOMISLAV LONGINOVIĆ, Contributing Editor of *Periplus*, was born in Belgrade, Yugoslavia, in 1955, and came to the United States in 1982, as a member of the University of Iowa's International Writing Program. He holds an MFA degree (fiction) from the Iowa Writers Workshop and a PhD in Comparative Literature, also from Iowa. Currently he is an Assistant Professor in the Slavic Department at the University of Wisconsin, Madison. His novel *A Moment of Silence* was published by Burning Books, San Francisco, in 1990.

ARVIND KRISHNA MEHROTRA, Co-Editor of *Periplus*, has published three volumes of poetry and one of translation, *The Absent Traveller: Prakrit Love Poetry* (1991). He is also the editor of *Twenty Indian Poems* (1990) and *Twelve Modern Indian Poets* (1992), both published by Oxford University Press. He lives in Allahabad.

GRZEGORZ MUSIAL, Contributing Editor of *Periplus*, was born in Bydgoszcz, Poland, in 1952. He is the author of five books of poetry and three novels, for which he has received major literary prizes in Poland. He published the first Polish translations of Allen Ginsberg's *Howl and Other Poems* and *Kaddish and Other Poems*. In 1982 he was imprisoned for his Solidarity activities. In 1988, he participated in the International Writing Program at the University of Iowa, and in the fall of 1989 returned to Iowa to work on anthology of contemporary American poetry in Polish.

LUCAS MYERS is a playwright, living in San Francisco with his wife, the Hungarian- born violinist Agnes Vadas Myers, with whom he translated the selected poems of Attila József.

KUNWAR NARAIN was born in 1927 and educated at Lucknow University, where he read English. His first collection of poems appeared in 1956, and it was followed by two others. He has also published a volume of stories and a short epic. Narain has been associated with several literary journals and cultural organizations, and is deeply interested in theatre and film. His many awards include the Premchand Award (1973) and the Kumaran Asan Award (1982).

SYBREN POLET, see headnote by James S. Holmes.

LIA PURPURA is a graduate of the University of Iowa Poetry Workshop. Her poems have appeared in a number of magazines. The translations of Katarzyna Borun-Jagodzinska presented here were made while she was attending the Translation Workshop at the University of Iowa, in 1990.

A.K. RAMANUJAN, Contributing Editor of *Periplus*, has taught at the University of Chicago since 1962, and is at present professor in the Department of South Asian Languages and Civilizations, the Department of Linguistics, and the Committee on Social Thought. In 1983, he was awarded a MacArthur Prize Fellowship, and was recently elected as a Fellow of the American Academy of Arts and Sciences. Ramanujan has published three collections of poems in English and two in Kannada. His volumes of translation include *Speaking of Siva* (1972), *Hymns for the Drowning* (1981), and *Poems of Love and War* (1985).

JON SILKIN is a leading British poet, who for many years has co-edited *Stand*, the literary magazine which he founded. His many collections of poetry include *The Psalms with their Spoils* (April 1980), *The Ship's Pasture* (July 1986), as well as *Selected Poems*, the most recent edition of which was published in 1988. Jon Silkin is the editor of *The Penguin Book of First World War Poetry* (1979) and the author of *Out of Battle: Poetry of the Great*

War (1972). He has lectured and read widely in the UK and abroad.

KITTY M. VAN LEUVEN-ZWART is Professor of Translation Studies at the University of Amsterdam. She organized an international colloquium on the legacy of James S. Holmes, at the University of Amsterdam, in December 1990.

DANIEL WEISSBORT, Co-Editor of *Periplus*, founded the London-based magazine, *Modern Poetry in Translation*, with Ted Hughes, in 1965. He subsequently emigrated to the United States, where he now directs the Translation Workshop at the University of Iowa. He has published three collections of his own poetry, and has edited a number of anthologies, including the Penguin *Post-War Russian Poetry* (1974) and *The Poetry of Survival* (1991). He has edited a collection of texts, *Translating Poetry: The Double Labyrinth*, which appeared in 1989.

JASON WILSON, Contributing Editor of *Periplus*, is Lecturer in Latin American Literature at University College, London. He has published two books on Octavio Paz, and essays and reviews on Latin American literature and culture.